Ground in Stone

Ground in Stone

Landscape, Social Identity, and Ritual Space on the High Plains

Elizabeth Lynch

LEXINGTON BOOKS
Lanham • Boulder • New York • London

Published by Lexington Books
An imprint of The Rowman & Littlefield Publishing Group, Inc.
4501 Forbes Boulevard, Suite 200, Lanham, Maryland 20706
www.rowman.com

86-90 Paul Street, London EC2A 4NE

British Library Cataloguing in Publication Information Available

Library of Congress Cataloging-in-Publication Data

Names: Lynch, Elizabeth, 1964- author.
Title: Ground in stone : landscape, social identity, and ritual space on the High Plains / Elizabeth Lynch.
Other titles: Landscape, social identity, and ritual space on the High Plains
Description: Lanham, Maryland : Lexington Books, [2021] | Includes bibliographical references.
Identifiers: LCCN 2021036069 (print) | LCCN 2021036070 (ebook) | ISBN 9781793618924 (cloth) | ISBN 9781793618931 (ebook)
Subjects: LCSH: Indians of North America—Chaquaqua Plateau (Colo. and N.M.)—Antiquities. | Bedrock ground stone implements—Chaquaqua Plateau (Colo. and N.M.) | Chaquaqua Plateau (Colo. and N.M.)—Antiquities.
Classification: LCC E78.C43 L96 2021 (print) | LCC E78.C43 (ebook) | DDC 978.8/9601—dc23
LC record available at https://lccn.loc.gov/2021036069
LC ebook record available at https://lccn.loc.gov/2021036070

∞™ The paper used in this publication meets the minimum requirements of American National Standard for Information Sciences—Permanence of Paper for Printed Library Materials, ANSI/NISO Z39.48-1992.

I dedicate this book to all stewards
of the land—past, present, and future.

Contents

Acknowledgments

RECOGNITION OF NATIVE AMERICAN LANDS

This book is about landscapes and ancient social spaces, so it is important to emphasize that all my research is carried out on the lands of displaced peoples. The bedrock ground stone (BGS) features discussed were created by the ancestors of living descendant Native American communities who recognize and claim the material culture as integral to their cultural heritage. Many of the spaces may be sacred. Although I am not a member of these descendent communities, I hope the work I do is presented in a good way that provides honor to all those who dwelled in or trekked through these majestic canyons of the Chaquaqua Plateau. My hope is that knowledge about these places can help heal the scars of injustice.

Foreword

Mary Lou Larson and Marcel Kornfeld

Bedrock ground stone (BGS) facilities are global phenomena with restricted ranges. Some regions explode with BGS; in others, these features are absent. California is a well-known focal area but by no means an isolated case. The sheer quantity of BGS in California has made it an occasional focus of study; however, its full potential for interpretation is far from attainment. Global summaries or syntheses included in this book are virtually non-existent, making Lynch's study unique, welcome, and a necessary contribution to the field. Although largely absent from the North American Great Plains, a small section of the High Plains abounds in BGS facilities. Generations of archaeologists since the 1930s noted BGS (often under different names) on the Chaquaqua Plateau of southeastern Colorado. However, they have not been theorized and systematically recorded until now! The present work is the culmination of 80 years of noting the existence and ten years of systematic research focused on BGS facilities of the Chaquaqua Plateau. More than little of the impetus driving the recent expansion of research is the initiation of salvage archaeology at Fort Carson since the mid-1970s. The early surveys and excavations were followed by the intensive recording of rock images, mostly by Larry Loendorf. Given that rock images are nearly exclusively at closed sites, the same locations sporting most prominent BGS facilities, as we learn in this volume, not a minor impetus behind the author's initiation of this project was Loendorf's council to do so. Partly because of the trajectory of initiation of BGS facility studies and partially the author's theoretical proclivity, the social and ritual perspective is strongly represented in this book.

Investigations of BGS on the Chaquaqua Plateau take us on a journey from the prehistory of the region, including the history of investigations focused on ground stone facilities, to the ethnographic and ethnoarchaeological synthesis of selected North American groups using BGS providing a broad context for these features, to an extensive review of BGS facilities on the plateau, in part I. Part II introduces JE Canyon, the focus of Lynch's study and systematic analytical methods followed by two case studies employing cutting edge photogrammetric and 3D visualization methods for capturing variation and nuances of BGS. Especially significant are the estimates of effort involved to create (intentionally or through use) the facilities based on experimental ground stone studies. Such estimates provide a basic line of evidence for understanding and interpreting the use of BGS. Although difficult to temporalize or indeed date BGS facilities, in part III, Lynch leads us through cultural time and landscape of Chaquaqua Plateau and into the most important part of her work, the ethnographic and historical frames of social configurations of grinding. Here Lynch clearly shines, bringing in theoretical perspectives on social life as well as narratives, including gender narratives to place grinding into the larger social and socio-reproductive sphere.

We have not learned all about BGS facilities globally or in southeast Colorado, but we have the first comprehensive synthesis of these facilities on the Chaquaqua Plateau. More significantly, however, is the employment of innovative analytical methods (photogrammetry and 3D visualization), but topping it all off are the social implications of BGS with global implications for future studies.

Preface

The archaeology of bedrock ground stone tools is a global endeavor. In North America, the most well-known studies are in California and the Great Basin culture areas, where archaeologists and anthropologists have had more than 100 years to develop regional perspectives. On the Plains, we are only just beginning to recognize the significance of these spaces to regional archaeology. The Chaquaqua Plateau is a unique area to begin this research.

I first stepped foot in southeastern Colorado as a field tech with New Mexico State University in 2003. Mark Owens and Larry Loendorf gave me, a non-traditional graduate student, the chance to learn about archaeology on the Plains, even though I had no prior field experience. The first time I saw the Purgatoire River from an overlook near our base camp, I was hooked. The work was hard. The hours were from dawn to dusk. The terrain was rough. None of the difficulties mattered.

Years later, when I spoke with Mark and Larry about getting a Ph.D., they suggested I write a dissertation on the regional bedrock "metates." Mary Lou Larson and the graduate admissions folks at the University of Wyoming agreed that someone needed to study these challenging, often problematic, components of past material culture. The Department of Anthropology funded most of my fieldwork through graduate assistantships and travel seed money. In addition, the College of Arts and Sciences through, a summer grant, funded a study of packrat middens near one of the BGS features. Finally, the 3D Visualization Center accepted me as an unofficial research associate so I could use their facilities and expertise to model the sites in 3D and Virtual Reality. Mary Lou Larson's guidance kept me on track. No one had systematically studied the Colorado features. They were not collectible, nor could you date them. The last obstacle did not deter me, although it probably should have. Archaeologists have to be able to tell time.

Nonetheless, Mary Lou Larson suggested that the best way to approach the features was through the broader theoretical lens of landscape archaeology. But where to begin? Luckily, Larry Loendorf had some room for me on a Colorado Rock Art Association project in 2008. That project opened up the opportunity to develop a landscape perspective of the features. His ability to understand the importance of rock art coupled with the proximity of rock art to many BGS features in the project area meant I could focus my research on developing robust analytical methods. It soon became apparent that the spaces were essential to human cultural processes, particularly in forming social identity and the reproduction of social norms and values. This book is only the first step in understanding how the process works.

So many people have encouraged and inspired my work; it is impossible to name them all. First, my family supported every aspect of this journey, from changing careers to going back to graduate school. My grandmothers taught me that social reproduction processes, no matter how theoretical, began in their gardens and kitchens. Summers spent on the farm with aunts and uncles, cousins, and friends during the harvests taught me the stories about everything that mattered to the community. They taught me who I was through the hard work of food processing. More than that, they showed me the process of becoming a good person.

Many people supported my thinking and imagination for this book. Mary Lou Larson's tireless efforts to help me craft the research were supported by my dissertation committee, Pam Innes, Marcel Kornfeld, Michael Harkin, Todd Surovell, and William Gribb. Conversations with my New Mexico State friends and colleagues, Mark, Scott, and Brenda, kept me on track while friends from a previous life Julie Allan in particular provided encouragement from afar. Anne McAbee, JoAnne Dreyer, and Clare Bowen O'Connor read countless drafts, helped bounce around ideas, and listened patiently through many trains of thought. Pam Innes and Marilyn Yee provided a writing haven through the COVID-19 pandemic. My writing group, Tia Frahm and Janel Seeley, provided invaluable support in the craft of productive writing. Finally, anonymous reviewers have provided invaluable insights that helped me improve the prose.

The archaeologists at Fort Carson and Pinon Canyon provided support in fieldwork, access to sites and files while also helping me contact Native American descendant communities. In particular, this book owes a considerable debt to Jennifer Kolise and Pam Miller. The Colorado

Office of Archaeology and Historic Preservation (OAHP) provided additional support. Mary Sullivan offered insight into the best search terms to capture the full scope of BGS features on the Chaquaqua Plateau, as well as vital expertise about the GIS files which supported the maps. Kallie Sanders was instrumental in gaining access to electronic site files as the pandemic began to materialize. John van Keuren, the Director of the Huerfano Heritage Center, provided much-needed background information on Janet Chatin. Kit Page, the Southwest District Manager with the State Land Board, assisted with access to Trinchera Cave. Dr. Christian Zier provided helpful information on the background at Trinchera Cave. Loretta Martin at the Louden-Henritze Archaeology Museum at Trinidad State Junior College helped with images and stories about the early days at Trinchera. Finally, I also owe a debt of gratitude to the Southern Ute, Arikara Pawnee, and Hopi cultural preservation officers for their support in determining the appropriate and potentially inappropriate presentation of images and data.

As with most endeavors, we stand on the shoulders of those who've come before us. I have done my best to provide well-rounded arguments and sound analysis with rational methods. Any errors or mistakes within these pages are my own shortcomings.

Introduction

Over the past few decades, archaeological investigations into bedrock ground stone (BGS) features have gradually become more established within the field. The growing interest in the science of ground stone tool analysis partly stems from the emergence of newer forms of scientific visualization such as 3D modeling and virtual reality (Lynch et al. 2017; Willis et al. 2016). Moreover, recognizing the socially symbolic aspects of these types of workspaces (for example, Claassen 2011, 2016; Lynch 2014) has sparked the public's imagination while incorporating Native American descendant voices in next-generation research projects.

The emergent area of BGS feature research employs numerous methodological approaches under various theoretical frameworks. In North America, technological studies, whether manual or digital, are fully rooted in seminal research methodologies (Adams 1999, 2002; Bartlett 1933; Eddy 1964; McCarthy et al. 1985; True 1993; Woodbury 1954). Developments in field recording methods led to breakthrough analysis (Leftwich 2010; Lynch 2014; McCarthy et al. 1985) that paved the way for building larger, comparative data sets such as those in chapter 2. Recent investigations pursue enhanced methods to understand how tool morphology and design are related to past human behavioral processes (Dubreuil and Savage 2014; Franklin et al. 2016; Leftwich 2010; Lynch et al., 2017). Alongside those research projects are the analyses of organic residues (for example: pollen, phytoliths, and lipids) to infer tool use or assess environmental variations (Buonasera 2012 and 2016; Pearsall 2000, 2019). This book addresses the materiality of bedrock ground stone features, a specific category of ground stone tools found in the canyons of southeastern Colorado. Their physical inaccessibility means these features have remained obscured from the archaeological gaze. The lack of methodological rigor in describing the regional features means they have been awaiting inclusion in the way we construct our knowledge of the past.

In archaeological knowledge production, one of our fundamental goals remains the development of meaningful information about the long-term dynamics of our changing humanity using what endures of the material objects, once part of lived cultural behavior. Some archaeologists are interested in how individuals and their societies construct worldviews or imbue life experiences with significance. To develop our interpretations, those working with cultures from time without oral or written records do so with incomplete information. Artifacts (portable cultural objects), features (non-portable cultural objects), and landscapes (natural, culturally-interactive environments), once part of active, living systems of behavior, now exist as part of a fragmented, palimpsest of materiality. We utilize two main tools to build a reasonable view of the past from the bricolage of this landscape: the physical leftovers of past activities (for example: tools, campsites, housing, or skeletal remains) and analogy to modern, living societies or historical information from descendant communities (Binford 1980; Schiffer 1983). Two studies that have proved helpful in the interpretation of bedrock ground stone features include the African ethnoarchaeological studies of David (1998) and Gardi and Gardi (1995). North American archaeologists also rely on regional approaches to interpret bedrock ground stone features. McCarthy and colleagues (1985) worked with Mono descendant communities to connect living memories and practices with the morphological analysis of bedrock mortars in California. Claassen (2016) provides multiple approaches to the socio-symbolic world of female workspaces for the bedrock mortar holes found throughout the southeastern United States.

Nevertheless, we are often working with limited data sets that can take us only so far. Information gathered about past societies must consider multiple lines of evidence from the analysis of material culture. By the latter, I mean tools, the remnants of tool construction processes, architecture ruins, or indications of habitation. Once developed, a reasoned image of the past emerges providing valuable knowledge about the human journey to our modern selves. An understanding that aides us in imagining our future. Since materiality may be fragmentary or non-existent, archaeologists often use experiments to help reconstruct reasonable connections about behavior and technology. An example of experimental research helpful to bedrock ground stone analysis is a use-wear study conducted by Wright (1993). Wright ground corn on a bedrock slab to determine the amount of time it would take to create depth from grinding on a new surface. Lynch (2017a) used Wright's result to posit a reasonable, although not literal, scenario for how long bedrock ground stone features could have been used before finally abandoned.

The most effective avenue for archaeologists into the nature of our human past is material culture analysis which considers the types of tools used but also how they may have changed through time and space to meet the needs of a group of people. For instance, projectile points shed light on the type of hunting strategies that may have been practiced. We can compare the development of projectile point technology across time periods throughout the world. The presence or absence of ground stone tools (*metates* and *manos*) in a site assemblage (the makeup of all material remains in a given location) is often associated with food processing. Thus, archaeologists often use ground stone tools to demonstrate foraging strategies for grains, tubers, nuts, or other local food resources. This type of inference relies on analogy to the lifestyles of modern hunters and gatherers (Binford 1980).

The tool types used for hunting (projectile points) and food processing (ground stone tools) in an assemblage may illustrate more differentiated subsistence strategies. Providing proper diets may have required multiple feedback networks encompassing all available human labor and knowledge as part of a more comprehensive environmental adaptation. Viewing archaeological assemblages from this context provides opportunities to explore questions about the mobility of past societies, how emergent horticulture lifestyles developed, and the division of labor, if any, within that society. In short, ground stone tools provide a means of understanding how people inhabited landscapes, not only how they moved but also how they utilized the resources they encountered during their tenure.

By definition, permanent ground stone tools are not movable, which is to say they remain in the locations where they were created and used. The stability of place provides an opportunity to develop in-depth knowledge about landscape use from a practical standpoint and a symbolic perspective. Claassen (2016) provides an example of permanent ground stone features in landscape context for the southeastern United States. In regions with documented use of BGS material culture, such as California and the North American Great Basin area, ethnography provides strong evidence for how people incorporate bedrock ground stone features in their daily lives (Barrett and Gifford 1933; Gayton 1948a, b; Gifford 1932, 1936, 1967). However, in other areas with copious BGS features but little or no ethnography, such as southeastern Colorado, understanding the material culture or building reasonable interpretations of their role in ancient [illegible] strategies turns out to be problematic. This book is different [illegible] that provide methodological appraisals for developing func- [illegible]eory (Adams 2002; McCarthy et al. 1985) about ground stone elements and features. Here I present multiple lines of evidence (low-level descriptive analysis coupled with comparative analysis) to develop an overall picture of prehistoric landscape use in the canyons of southeastern Colorado, where BGS features are widespread. The book develops our understanding of how people living in the canyons may have used these types of features. The results contribute to global comparative data so we can understand the socio-cultural implications of grinding practices worldwide.

WHAT IS MEANT BY BEDROCK GROUND STONE FEATURES?

The central premise is that the bedrock ground stone (BGS) features were an *integral* part of the High Plains culture. As such, they should be included in any discussions about culture on the North American plains before Euro-American colonization. The geographic area discussed in this book is limited to southeastern Colorado. The field expeditions of E. B. Renaud in the 1930s and 1940s describe BGS features in neighboring areas of New Mexico, Texas, and Oklahoma. Considering the potential range of cultural variation across the High Plains, this book focuses on features discovered on the Chaquaqua Plateau, a landscape designation (Fenneman 1931) confined to the southeastern corner of Colorado. The study enlarges investigations established in a single side canyon within the plateau (Lynch 2014). The earlier research provided a dedicated view of BGS feature variation within a geographically bounded but interconnected landscape on the JE Canyon Ranch. The limited study area allowed the author to focus on and describe the range of human workspaces in a defined geographic space. Methods developed during JE Canyon research combined with an expanded database from the Colorado Office of Archaeology and Historic Preservation (OAHP) provide a frame of reference for the BGS features recorded throughout southeastern Colorado.

The Chaquaqua Plateau is a relatively diverse geographical unit consisting of canyons, plains, and mountain or foothill ecozones. People have inhabited this region for at least 10,000 years, although there may have been periods during which the population sizes varied due to periodic climate changes (Cassells 1997; Stone, 1999; Scheiber and Clark 2008; Wedel 1963; Zier et al. 1999). However, it seems that the canyon system may have provided an advantage to weathering such climate shifts (Michlovic 1986). Over time, beginning about 5,000 years ago, people seemed to have developed lifeways attuned to the canyon-mountain-plains ecosystems. This sustainable self-reliance may have buffered interactions with neighboring cultures that developed traditional agricultural practices. The addition of BGS feature analysis to regional studies increases the known range of morphological variation, which, in turn, enriches our understanding of the regional cultural practices through time.

BGS features may represent the primary permanent material remains of a technological system for a seemingly resilient food-getting-processing tradition. The features also embody social processes that sustained local communities through ritual practice and socio-cultural reproduction. However, unearthing the differences between features used for expedient food or resource extraction and those used as part of long-term social practices requires a regional and comparative perspective.

Most data on BGS features in southeastern Colorado derive from cultural resource management work. In the field paperwork and subsequent reports, BGS features tend to be consolidated together under the umbrella heading of 'vegetal food processing' or 'faunal and flora processing' to the detriment of an inclusive construction of the past. BGS features were an essential part of food-getting-processing strategies or networks as well as lifestyle choices; therefore, they should be considered part of the entire archaeological assemblages in the region. As part of everyday practices and social processes, the features anchor human engagement with the many local environments found within the Chaquaqua Plateau's canyon systems. BGS features may hold the key to understanding cultural resilience within the region, with subsequent implications to relationships built with people in neighboring regions through the interconnected hydrology systems of the Arkansas, Canadian, and Cimarron Rivers.

BGS features throughout North America are found in tracts of forests producing predictable harvests of nuts and fruits (for example: oak and pinon). They are also found in resource niches with a wide variety of predictable plant species such as the juniper-woodland or juniper-grassland environments familiar on the Chaquaqua Plateau. The features are generally near good sources of permanent water, which include local springs. However, those found on the Chaquaqua Plateau show a wide variation in locations. Some are in canyons; others are in open-air contexts that may be far from water sources. The BGS feature morphology, which varies depending on the landscape context, provides a means to further explore these features' life history.

Archaeologists use many names to categorize regional BGS features. As part of archaeological assemblages, the answer to *how* food or other resources were processed is assumed. We infer the types of material processed are local plants unless corn is found in association with the BGS feature. The inference creates a potential source of bias. It assumes only food processing occurred on-site, while variation in shape might indicate that some features were used to refine tools or make hides. It also ignores whether the features could have been multi-purpose tools for making pottery, extracting mineral resources in addition to food-processing (e.g., Adams 1999; David 1998; Gardi and Gardi 1995; Woodbury 1954).

Additionally, *how* features were used as part of mobility strategies is an essential consideration. What does variation in shapes or types of BGS features tell us about land-use patterns? Since this book aims to weave the existing features into regional archaeology, understanding the variation of the types of features called "bedrock features" is discussed in some detail, along with methods that are helpful for field recording and description.

Two familiar names applied to BGS features are the bedrock mortar (a term common to California and the Great Basin archaeology) and the bedrock metate. Bedrock mortar features are also found throughout southwest Texas and northern Mexico. In the southeast, similar features are called bedrock mortar holes. "Bedrock mortar hole" is rarely used in southeastern Colorado archaeology. The second catchall term, bedrock metate, generally refers to all grinding surfaces oval-shaped with a well-defined basin. In some field records, the term bedrock slick or basin has been used for the bedrock metate. The difference between a bedrock slick and a bedrock basin or metate is that the term "slick" is an amorphous or irregularly shaped surface with little or no depth. These terms are discussed in detail throughout the book, but a brief introduction is warranted here. The bedrock mortar is characterized by a fairly uniform circular perimeter or rim. Generally, these features are deep (around 30 cm) and are associated with a regular pounding motion using a large pestle (stone or wood) to crush the food resources. The only way to understand the applicability of terms used by early archaeologists to describe these features in Colorado is to compare morphology with well-defined shapes from other regions (chapter 2).

The BGS feature (whether they have one or many grinding surfaces) is considered separately from grinding stones that are quarried, then shaped for use within households or community plazas. BGS features occur on naturally occurring landscape features such as boulders or erosional escarpments. They must be traveled to for work. Some sites consist of only a single BGS feature with one grinding surface. Without any immediate cultural context, such as encampments or long-term habitation areas, we must rely on naturally occurring plants and features or frequency of other types of sites in the general vicinity. This situation appears to be especially true of the BGS features found throughout southeastern Colorado. Consistent terminology is critical to developing a regional picture of bedrock ground stone features. Morphologies and site types illustrate the regional variation, not terms such as "bedrock metate," "bedrock mortar," "bedrock slick," or any variation of these terms. Instead, the term bedrock ground stone (BGS) feature is used throughout. Terminology, including variation in the application of descriptive terms, will be unpacked in chapter 2.

For introductory purposes, the BGS feature is generally a large chunk of bedrock that has been exposed due to erosion of sediments or boulders that have cleaved from a parent rock within rockshelters. Regional variations exist in the types of parent rock material. Sites are found in many natural contexts, including rockshelters and open-air. The people on the High Plains preferred the readily available Dakota Sandstone as their primary material. Dakota Sandstone occurs predominantly along the canyon rims and terraces, where most of the rockshelters are found. Many BGS features were created on exposed bedrock escarpments in the grasslands above the canyons. The latter provides the opportunity to discuss feature identification and differences in practical use and social processes.

Although Paleoindian peoples may have created some BGS features, the bulk of the evidence points to their development during the Archaic about 3,000 to 5,000 years ago, increasing popularity until the end of the Late Prehistoric about 500 years ago. Diagnostic materials such as projectile points, ceramics, and architecture are combined with radiocarbon dates, found in proximity to some BGS features, to discuss their presence through time. Two main trends appear along the Chacuaco: the creation of basin-like depressions on the large, sheltered boulder surfaces and the much more widespread BGS features found in open-air assemblages on eroding escarpments.

In the earliest field reports, archaeologists noted the similarity of local BGS features to the metate-mano grinding tool kit standard among the Pueblo, Zuni, and Hopi peoples living in the neighboring areas to the south and southwest. They also notice the similarity to grinding implements used by the Navajo peoples, whose long history intersected with the southeastern area of Colorado sometime in the fifteenth century (Gunnerson 1989). The analogy between BGS features and the Southwestern mealing bins became a reasonable way for archaeologists to think about the regional features. The assumptions affected how BGS features were incorporated into the general prehistory of the region, which is to say that the features were often mentioned only in passing as evidence of shifting reliance on a corn-growing lifestyle. The assumptions made by earlier archaeologists were so firmly ingrained with the standard ground stone tool kit's image that recording of BGS features was often overlooked, so few were documented in detail until cultural resource management work in the 1980s. First-wave archaeologists might even question the need for a book dedicated to the BGS features found throughout southeastern Colorado.

PROBLEMS

One of the main arguments interfering with the inclusion of BGS features in the technological system for a resilient food-getting tradition revolves around the existing evidence. A question often posed is whether we can know what role the BGS features played in past societies, especially those ubiquitous open-air features that may have been short-term seasonal workspaces. As yet, there is no direct link between BGS features and modern descendant communities from the High Plains, but indirect knowledge prevails. Therefore, the only thing we can know about the features is that they were used for grinding something in the past. As a result, many rely on the following assumptions about BGS features: 1) they are part of a transition to pre-agricultural lifestyles, 2) they represent women grinding food, or 3) they are an expedient means of supplementing a diet based primarily on meat resources for highly mobile societies made up of small groups of people (around 25).

Further, the argument goes, there are too many impediments to dating the features. To tie BGS features to a "central" or "integral" part of prehistory, we must be able to place them into the larger context of time. While we can discuss their spatial distribution and will, we have to construct a foundation to model how BGS features may have functioned along with other parts of the site assemblage. This book employs an ethnographic comparison of grinding cultures to illustrate the critical role such features played in subsistence strategies. Chronology is reconstructed by compiling dates from associated archaeological assemblages, including projectile points, ceramics, architecture, radiocarbon dates, and rock art seriation. Together, the lines of evidence provide a relatively solid base from which to launch an inclusive regional evaluation of BGS features.

A separate argument opposed to this book's central premise originates in the skepticism that we can never really understand past social processes in archaeological material culture. Even if we had direct ethnographic evidence, we could not know how much has changed due to colonization or other influences. Under the best of circumstances, the deduction of social processes using the remains of material culture may represent post-modern mumbo-jumbo, even if one can tease out the palimpsest of human-human objects-nature interaction through time. In southeastern Colorado, we lack direct ethnographic links, and we have a problem integrating the BGS features with other artifacts and features. Nonetheless, grinding toolkits and practices from other regions provide a robust picture of how grinding merges with social rituals. They also illustrate that grinding space is a central location where much of human identity is created, maintained, and passed on to later generations.

METHODS AND INTERPRETATIONS

The regional inclusion of BGS features is a problem that requires multiple approaches, from morphometric analysis to oral tradition. The existing interpretive framework needs re-evaluation in light of tools from all anthropological subfields, including a wide array of ethnographic data but also the scientific collection of archaeological data that can be rigorously tested and replicated. A dialogue has to be created between the scientific process and descendant community knowledge systems.

The recent developments in 3D visualization products provide us with multiple ways to record bedrock ground stone assemblages supporting traditional data gathering methods. Changes in problem visualization can occur early in the research process, thus aiding us in creating better research designs and improved hypothesis testing. While 3D visualization methods may not completely replace standard field recording, they do significantly reduce the error in manual field recording while increasing the replicability of data. Manual recording processes of BGS features are time-consuming. Advances in 3D visualization allow us to process heavy amounts of data to describe features. Improved feature description supports an integrated understanding of the surrounding archaeological landscape. Both provide a dialogic method further to understand BGS features within the context of regional archaeology.

Reevaluating our current interpretive frameworks includes developing the precise descriptions discussed above but also incorporating Native American voices. In North America, oral traditions from California, the Great Basin, and the Southwest provide evidence that social identity is created at centralized grinding places. For instance, stories describe the ritual associated with grinding corn or pounding acorns (discussed more in chapter 8), thus providing examples of ritual practices concentrated on grinding activities. As a part of rites of passages, the grinding process or ceremony creates and manages the social identities of the participants. Grinding activities sustain community identities because of the rich knowledge of resources and long-term land tenure that are part of the subsistence dynamic. The Bear and Fawn story tradition illustrates the dynamic. Perhaps one of the most widespread narratives across language families in the western United States, the stories encode landscape knowledge and land tenure experience.

Many regional archaeological problems could benefit from the inclusion of BGS features. The scientific description of BGS features provides the foundation to incorporate the features into regional archaeological problems. One such problem is the archaeology of stone enclosures in southeastern Colorado. The landscape of these features consists of a mosaic system of canyons, foothills, and plains ecosystems. E. B. Renaud (1930) speculated that the position of local slab stone enclosures on higher terraces supported an early warning system for people living in the canyons. Although Renaud and others have focused on the need to mitigate human aggression or provide hunting outlooks, the sites would also have been an effective warning system for natural disasters such as fire and flash flooding. Some stone enclosure sites neighbor BGS features (Gunnerson 1985), but the features were not recorded. If the canyon system lifestyle was as long-lived as we suspect, people living within the area would have had to manage such environmental perturbations successfully. Incorporating analysis of BGS features that neighbor slab stone enclosures provide insight into these discussions.

The organization of this book follows three broad themes. The first develops an overview of the research challenges involved with the study of BGS features. Chapter 1 looks at the history of archaeological investigations into BGS features for the region. BGS features seem to be invisible to the archaeological gaze, but early archaeologists recognized them in the field and sometimes recorded the locations of such features as they encountered them. However, because early archaeologists were interested in other questions, such as the colonization of the Americas, archaeological contexts were developed on those artifacts most easily collected and analyzed to address those problems. It may be that a researcher's assumptions about the nature of the BGS features (for example, that they were essentially synonymous with the grinding implements in the neighboring regions) resulted in conclusions that little work needed to be done. It may have been that early archaeologists unconsciously minimized the hypothesized female labor, thus effectively relegating them to the sidelines of prehistory. The reasons are of interest now as only a means of understanding the historical trends in regional archaeology. Nonetheless, these early archaeologists provide some insights into the distribution of the BGS features and their relationship to other sites. Their work illustrates how the BGS features were interpreted long before the techniques were available to describe them fully.

Chapter 2 introduces the terminology and descriptive issues involved with BGS features. Although data are being collected to measure the different BGS feature elements (the individual grinding surfaces on each BGS feature), the contribution of the measurements to regional archaeology is often overlooked or seen as the rather tedious collection of unreplicable data. Chapter 2 addresses this problem by comparing data collected during a survey of JE Canyon in 2011 with those data verified by ethnographic consultation in California. JE Canyon data was published in peer-reviewed media. One of the main concerns with the regional archaeology of BGS features is the wholesale use of terminology that may bias interpretations. For example, metates and manos are often associated with corn grinding processes. Thus, the term "bedrock metate" alludes to functional implications that may not be true in southeastern Colorado. The generic application of such terms presupposes functionality before any analysis has been conducted. Chapter 2 poses a few critical questions. Are the BGS features on the Chaquaqua Plateau similar to those found elsewhere? Can we compare the shapes of the different grinding elements to understand regional functionality better? How can we improve the use of descriptive terms without applying colonizing vocabulary to a material culture for which we know so little?

Chapter 3 explores the spatial distribution for all the BGS features within southeastern Colorado. In this chapter, the locations where we are most likely to run into BGS features are described based on sites that have been reported to OAHP. The distribution of BGS features and their relationship to natural resources illustrate where the bulk of archaeological research has been focused but also offers some knowledge about landscape use.

Chapter 3 also examines the distribution of BGS features with other features, such as artifact clusters. The analysis explores how BGS features relate to the different landscape and archaeological contexts. Portable ground stone is part of the regional artifact inventory but is considered separate from BGS features. Many of the societies who lived in the canyon systems were adept with portable ground stone tools after being abandoned by those using the BGS features. Prehistoric and historical use of these tools, coupled with their portable nature, requires us to be conservative in our assumptions about the relationship between portable tool kits and the permanent BGS features. Other questions explored in chapter 3 include how the landscape affects the type of BGS features, especially the relationship between water and plant resources nearest to the BGS features.

The second theme illustrates the methodologies used to record the BGS features in the field. Using a case study approach, I show the evolution of the manual recording method that evolved from methods used to record bedrock mortars in California and portable groundstone tools in the Southwest. Chapter 4 discusses the traditional mapping and metric data collection methods applied in the field. The influence of different metric techniques is discussed as a single field case from the JE Canyon Ranch survey. Since these data have been published (Lynch 2017a, b), chapter 4 is concerned with developing the rationale of data collection for BGS features to help other researchers decide how to best design their data collection methods. Chapter 4 explores underlying assumptions about what we hope to measure with the various metrics and the information we hope to gain. I also discuss the cost-effectiveness of such techniques in conjunction with developments in digital data collection methods.

Chapter 5 illustrates developments in the digital collection and presentation of metric data using photogrammetry. The case study discusses work completed at Metate Cave (5LA6575) on the Pinon Canyon Maneuver Site (PCMS) east of Trinidad, Colorado. It is one of the more protected rockshelters. Analysis of rock art suggests a long-term occupation from the Late Archaic to the modern Euro-American presence. The site has a large number of grinding elements combined with rock art panels. Rock art superpositioned over existing ground stone basins or basins ground over rock art motifs suggests a dynamic situation that does not fit neatly within existing frameworks for regional BGS features. In addition to providing a full scientific visualization model for analysis, photogrammetry at the site also helps preserve the features for cultural resource management efforts, including consultation with descendant communities.

Chapter 6 explores the recording of Trinchera Cave BGS features and landscape. The site is critical to understanding regional cultural interactions; however, the BGS features have long been overlooked. One of the questions posed in this chapter is why we need to use and teach the application of new technology? The manual techniques are well-known and documented. If we hope to engage a broader audience in BGS research, there needs to be a means of replicating findings between studies within and across regions. Unfortunately, manual recording techniques have a steep learning curve, so they retain a high interobserver bias. The case study illustrates that creating accessible digital models (data sets) reduces bias and increases our ability to replicate the work of others. At the same time, we minimize traffic to the sites while preserving them for future research and study. Additionally, we can produce applications or 3D print models for museum exhibits which helps to enhance public outreach.

The third theme provides analyses that move BGS features beyond descriptive research by providing further examples to integrate them into the regional archaeological contexts. Chapter 7 uses existing dating methods to refine the timeframe of BGS feature use and begin to integrate them into the regional chronology. Our present understanding is developed from a few sites from dissertation research (Campbell 1969; Lynch 2014) or cultural resource management reports. This chapter combines evidence from all of the recorded sites to refine this chronology.

Chapter 8 develops the hypothesis that the creation of social identity plays a fundamental role at sites with BGS features. The chapter employs ethnographic examples of material culture and social processes related to grinding practices in the western part of North America. The chapter provides the foundation for an interpretative framework to approach the interconnected roles of social identity, technology, and subsistence practices. The emphasis becomes the cultural practices (food processing, ritual, and food types) at permanently constructed grinding features such as bedrock mortars or mealing bins. In addition to the subsistence practices, oral tradition is explored to encode symbolic and sustainable practice through repetition, imagery, and ritual practices.

Chapter 9 synthesizes the methods, theory, and fieldwork discussed throughout the book. It discusses the problems overcome so far and those that remain. This chapter proposes a research trajectory for the near future, including how regional avocational archaeologists, always one of the strongest pillars of regional research, can contribute to and help build a sustainable outreach system for BGS feature research.

Archaeological research into BGS features has been most vigorous in the western United States, although archaeologists in the eastern regions are pushing us to broaden our interpretive framework. This book reconciles the two regions by emphasizing the field data collections and interpretative frames of reference to develop the visibility of the Plains bedrock grinding phenomena. The hope is that this book contributes to the overall visibility of ground stone studies in the United States by providing innovative approaches to fieldwork and data visualization while also helping to bridge the archaeological understanding of High Plains landscape use.

I

INTRODUCTION TO THE RESEARCH PROBLEM

1

History of BGS Feature Archaeology in Southeastern Colorado

Early regional archaeologists mentioned bedrock ground stone (BGS) features in the Chaquaqua Plateau as early as the 1930s. Their notes were primarily descriptive renderings of the BGS features as part of their broader archaeological explorations of the local canyons. The feature descriptions by these early archaeologists rarely provide measurable data or analysis. The oversight may have been the result of the prevalent androcentric bias in the early days of archaeology. However, the situation was probably much more complicated because archaeology, at the time, was a relatively new scientific domain with evolving methodologies. Early archaeologists, navigating these uncharted and somewhat treacherous waters, learned quickly from mistakes to correct their course as soon as possible. Nonetheless, BGS feature analysis did not seem germane to the central question of interest to the scientific community of the day, namely the length of time humanity had been in the Americas, particularly North America, with corollary inquiries into how the continent had been colonized and by whom (Meltzer 1983). Wormington (1957) reasoned that the ancestors of modern Native American communities had populated the lands during the end of the last significant ice age, about 12,000 years ago, while BGS features related to grinding practices that developed at least 6,000 years later.

In addition to looking for clues into the earliest colonization of North America, another driving force for archaeological discovery in southeastern Colorado was the origin and spread of agriculture. Some of the earliest exploration for scientific discoveries included John Wesley Powell's expeditions in the 1860s, Jesse Fewkes's work among the Zuni and Hopi in the 1880s, Kidder's groundbreaking archaeological investigations in the southwestern United States (1890s), and Wedel's early 1900s research on the North American Plains. These works incorporated the material culture of living Native American communities into their exploration of archaeological questions, including those on the adoption of agriculture. While Puebloan mealing bins played a part in these studies, BGS features, in general, did not because they were used to process locally harvested resources (Powers 1882, Powers and Powell 1877).

Soon students of Franz Boas, considered by many to be the founder of the modern American field of anthropology, swarmed across the United States collecting information on Native American language, society, and material remains. They documented what they could, as cultural treasures were rapidly vanishing into obscurity. They focused on collecting and describing cultural data, which in many ways predicted the ways archaeologists would conduct themselves in the field: survey, dig, collect, and describe (Willey and Sabloff 1993). Onto this frenetic early stage of American archaeology walked perhaps one of the more remarkable archaeologists to work in the region. E. B. Renaud connected with the local imagination while keeping an eye on the more significant issues of his time. His relationships with local ranchers, schoolteachers, and other local adventurers gained him entry to some of the more, at the time, inaccessible areas of the region.

Etienne B. Renaud (1880–1973) emigrated to the United States after earning his degree at the University of Paris (1905). He became a US citizen in 1913, working in language training for the US Army during both world wars. His academic career at the University of Denver began in 1916 when he changed from a professor of Romance Languages to Anthropology. His influence on the emerging field of anthropology affected a new generation of Colorado archaeologists (Cassells 1997). Beginning in 1930, Renaud started his formative survey of the archaeology of the Chaquaqua Plateau region. His expeditions relied heavily on collaboration with local ranchers and citizens (Figure 1.1). His trips would follow the hydrological systems, taking him into the panhandles of Texas and Oklahoma, the northern expanses of New Mexico, then north to Wyoming.

Although criticized by some contemporaries (Dick 1946; Huscher 1943; Van Royen 1936) as somewhat speculative and quotidian, Renaud's work stands out for a couple of reasons. In the first place, the tracks of his expeditions follow the most accessible navigable routes through the regions. During the 1930s, major road systems which followed earlier trails and waterways were scant. Renaud describes the difficulty in accessing the sites without expert local knowledge, echoed in Haldon Chase's expeditions in the region during the late 1940s and 1950s. Even with "modern" maps, most of the sites were difficult to locate for outsiders. Thus, Renaud's journeys still provide the foundation for our understanding of regional BGS feature distribution.

Interesting archaeological sites were mapped onto the local knowledge as picnicking areas, hiking spots, and romantic places. Cattle visited many locations because of edible resources, water, or shelter, but locals visited these locales for various reasons. Along with local knowledge, Renaud tapped into the residents' passion for history and cultural preservation. More than that, he was instrumental in training landowners, schoolteachers, and others as avocational archaeologists, nudging their innate curiosity towards cultural heritage stewardship and away from antiquities collection. An example of this would be his long-term relationship with Janet Chatin, a Walsenburg schoolteacher turned avocational archaeologist, who conducted excavations on Renaud's behalf.[1] Aside from building a descriptive archaeological baseline for the region, Renaud was captivated by the resemblance of the local "stone enclosures" to European prehistoric stone monuments (Renaud 1942). He drew direct correlations between the antiquity and development of cultural sequences in southeastern Colorado based on the architectural similarities between the stone enclosures and European material culture. His methods are still applicable to modern research. Although it has received little mainstream attention, his quest to understand regional interaction has endured. Renaud's work still influences regional archaeology even though many of his chronological sequences have been replaced.

Renaud's influence anticipated how BGS features would be incorporated into the regional prehistory. In his 1930 expeditions of the region, Renaud (1930:139) writes of Cave 5, in northeastern New Mexico, "I have mentioned four metates from the upper fireplace and another from a deeper layer; there were also trough-like depressions in rocks near the entrance of the cave; all had been pecked in the center ovally as noticed for the Oklahoma grinding stones. Twenty-one-hand manos . . . were collected." The description of the food found in Cave 1 near Kenton, Oklahoma, eclipsed the grinding stones descriptions (Renaud 1930:124).

> Besides a variety of wild seeds, acorn, wild fruit, and berries, squash represented by seeds and pieces of rind, we have the absolute certainty that maize was cultivated. It is represented by corn husks, many grains of a single kind, reddish-brown in color, scattered all over; one ear with 10 spiral rows; 35 cobs; and 8 or 10 fragments, many very short and thin, suggesting a primitive type. We also found two grass bundles containing shelled corn and weighing 11 ½ ounces each, and a prairie dog-skin bag containing ears of maize and weighing 12 ½ ounces. Hence these ancient folks were early agriculturists. They also made food cakes of acorn mixed with wild plum or cherries. Three of these cakes, dark brown, circular, and now excessively hard, were recovered from Cave 1. They weigh respectively 11, 20, and 21.6 ounces. Their

> diameters range from 109 to 156 mm and their thickness from 80 to 35 mm. A small central hole was used for suspension by means of a string of grass fiber, probably to keep the cakes away from the rodents.

The description is important in a couple of ways. First, it sets the framework from which ground stone will be viewed—always in conjunction with food resources, but in particular with agriculture and corn processing. Although it seems natural enough to equate grinding tools with food production, especially food found nearby (in the above case, corn and acorn cakes), ground stone remains multifunctional tools that may be used for a variety of production processes (Adams 1999; Woodbury 1954).

Renaud was heavily influenced by AV Kidder's *Southwestern Archaeology*, newly published in 1924 (Renaud 1925), especially since so little was known about the southeastern region of Colorado, which seemed to be influenced by the Southwest and Southeast cultural areas (Figure 1.2). Archaeologists would have been influenced by the relatively recent discovery of the Folsom site in northern New Mexico. Renaud became dedicated to looking for evidence of "Folsom Man" and the expansion of agriculture in the region. Having worked on projects in the southwest (Cassell 1997), Renaud's understanding of ground stone terminology would have developed from the Southwestern corn-growing tool kit (mano and metate) as well as the *meules à grains* found in European archaeological assemblages.

Katherine Bartlett, a prominent Southwestern archaeologist, recognized that ground stone assemblages were all but ignored in the ongoing archaeological work in the Southwest and Plains. She suggested that their invisibility was due to their overwhelming *presence*. Bartlett (1933:1) posed the question, "Is it not strange that the corn mills of the ancient Southwesterners have received so little attention or thought? In a corn culture such as the Pueblos had, where their very life depended on their crops, what was the most important thing in their homes, if not their grinding stones? Without them their corn would have been of little use." Bartlett (1933) developed a framework to view ground stone within the ethnographic and archaeological contexts, producing a timeline for the development of grinding stones in relationship to architectural and demographic changes in Puebloan cultures.

Renaud's understanding of the bedrock grinding features seemed to develop beyond linking local material culture to adjoining agricultural complexes. This line of inquiry took a back seat to his other interests, namely the local stone enclosures and their relationship to the neighboring culture areas. However, his interest is evident in Renaud and Chatin (1943:19) where they describe a site designated C. 919A:

> At the east end of the rock shelter there is an enclosure, three paces deep by four paces across, among tumbled down boulders. On one of these rocks

> there are five grinding troughs: two of these "fixed metates" are deep, showing pecking and much use, the others are shallow. There are petroglyphs, of the pecked type, on the "metate boulder," but too weathered to allow deciphering except for a whirlwind. There are as many troughs on another rock, and a large fire place between these two boulders near the middle of the opening of the rock shelter. . . . There is a better spring near the trail just before reaching the rock shelter.

The cognitive influence of Southwestern material culture is evident in the use of the term "trough" to describe the ground stone elements. These are ground stone types common in the four-corners region, even on bedrock (Villalobos-Boehm 1997) but are rare in the BGS features of the Chaquaqua Plateau. Additionally, using the term "fixed metates" is an apparent reference to the fixed mealing bins, which are a dominant aspect of Puebloan material cultures.

Renaud and Chatin (1943:26) discuss material remains from another site (C. 922):

> [V]ast rock shelter faces north and its view is blocked by the cliff on the opposite side of the canyon . . . on the boulders forming the fences on each side of the enclosure 'a' one sees two shallow depressions which have been used like fixed metates. A large, heavy moveable metate with a very deep trough, clearly pecked for grinding, was also found in the same squarish compartment 'a,' and a heavy eight-inch stone which might have been used as a pounder, one end being smaller could have been held in the hand. In the back wall of enclosure 'b,' there are three tool grooves.

Janet Chatin excavated, per Renaud's request, "near the grinding stones" in the locality they named "a." Her description provides evidence for local resource grinding, not corn flour production.

> [B]elow the loose surface dirt, there was a layer of sand and charcoal containing a large, heavy pecked two-hand mano. In screening the dirt, small corn cobs, charcoal, broken bones, a pinon cone, some small seeds, and wild plum stones were found. Then there was a bed of grass lying in a horizontal position and apparently extending beyond the pit. A pointed piece of bone, useable as a boring tool, was found near the bottom of this pit, some two feet below the surface. Renaud and Chatin (1943:47)

Renaud and Chatin's (1943:47) interpretation of C.922:

> In one compartment fire places and bones clearly suggest the kitchen or cooking place; on the wall of the next cell tool groove and the flakes picked from the floor might indicate the workroom or workshop; in the following section the presence of a large moveable metate and of two fixed metates or grinding troughs make it clear that this was the milling or grinding room

> much as is seen in Pueblo houses; finally at the south end of the shelter and again between the wall and transversally standing rock we see four groups of pictographs as if this part of the cliff village or dwelling was reserved for ceremonial purpose.

Early archaeological work assumes grinding processes mirror those found in the southwestern United States on one area of the site. There are no measurements of the use surfaces, so the morphology of the grinding surfaces is not replicable. Finally, in this interpretation, we see the new trend emerging in which ground stone features are linked, through the presence of rock art, ceremonial food production, or, at a minimum, the connection between food production (generally through female labor) and ritual space. At another site in the Cucharas Drainage, Renaud and Chatin (1943:360) describe the features at another site (C 935):

> The rock shelter measures eight feet in height, sloping back to four feet. The wall is smoked and small boulders are scattered near both ends of the shelter. There are fixed metates of shallow grinding troughs on two large rocks, one 3x4 feet, the other 2.5x4 feet, near the north end of the shelter . . . a small test hole was dug near the "grinding rocks"—found charcoal.

Renaud and Chatin continue to describe the material culture in the context of the neighboring Puebloan culture. The similarity of the ground stone tools with fixed mealing bins and corn horticulture set the stage for future interpretation of BGS features in the region. This influence meant thinking about the BGS features in terms of more sedentary subsistence patterns, conjuring up images of groups of women working together to grind corn for ceremonies or daily family needs (Table 1.1).

The image is so fixed in the minds of local archaeologists that the terms used to describe the features themselves perpetuate the ideal. We will unpack the terminology more in the next chapter. Chapter 8 will explore the implications of this bias to our understanding of social identity and practice by the ancient peoples who created and used these features.

The stone enclosures became a particular problem for Renaud as he traveled the Plains. In his descriptions, these enclosures were very close to European structures capturing the public's interest in places like Stonehenge. Renaud also compared them to *menhirs*. He recognized the importance probably placed on these structures by the ancient regional inhabitants. He also began looking at how they were placed along the canyon rims and in what manner they were related to other features found nearby. The relationship between these stone enclosures and BGS features has yet to be addressed in regional work. However, in chapter 3, we will look at the relationship between local architectural features (such as stone enclosures) and BGS features.

Table 1.1. Summary of Renaud and Chatin (1943) of the Material Culture Associated with BGS Features

Site Designation	*Context*	*Ground Stone*	*Rock Art*	*Other Material Culture*	*Implications for Social Identity and Structure*
C.919 A	Rockshelter	Grinding troughs Fixed metates Metate boulder	Pecked rock on boulder "whirlwind"	Rock wall, Fireplace between boulders Spring nearby	Cooperative group work Female labor collective Possible ceremonial connection Rock Art
C.922	Rockshelter	Fixed metates Large portable trough metate 8-inch stone pounder?	Tool grooves	From excavation (at least two feet deep) near "grinding stones": Sand and charcoal Two handed mano Corn cobs Broken bones Pinon cone Small seeds Small corn cob Bed of grass Piece of bone (boring tool)	Cooperative group work Female labor collective Cooking and food preparation Possible tool making area Possible ceremonial context related to Rock Art and tool grooves
C. 935	Rockshelter	Fixed metates on boulders	NA	Charcoal in test pit	Cooperative group work Female labor collective Cooking/food preparation

Reviewers often noted Renaud's preoccupation with thunderstorms on the horizon. To those living in the canyons of the plateau, flash floods are a real threat, frequently recreating the floodplain landscape. Local inhabitants and archaeologists are hyper-aware of developing weather systems throughout the region, both in terms of generating fires from lightning strikes and flash flooding. Although Renaud thought the stone enclosures related to defensive behaviors, his concern with weather systems conjures up another scenario, in which the stone features may have served as early warning systems for flash flooding or other types of environmental disturbances.

Another critique of Renaud's descriptions is the inability to relocate some of his sites. Gunnerson (1989) traced well-known Apishapa structures to fieldwork described in Renaud's fieldnotes and publications. However, many of the sites describing BGS features may never be recovered. Regardless, Renaud placed value on general descriptions of the grinding features even though he seemed uninterested in developing methods to analyze them or their place in the regional context. His excavations at the base of BGS features suggest that he believed the evidence of resource production would be there, though he seemed to persist in believing they were related to corn-growing practices.

These early assumptions became entrenched in future interpretations that the BGS features were related to shifts from hunting and gathering lifestyle to more sedentary horticultural pursuits. For example, the heavier the use of features (determined by the number of ground stone elements and their depth), the more likely people using them practiced horticulture. Ultimately, Renaud was unable to place the BGS features within a regional or multi-regional chronology. Few took up Katherine Bartlett's challenge to better study ground stone, as most continued to focus on the search for early human occupation, the spread of agriculture, and the more pressing problem of developing a coherent chronology for Plains archaeology.

REGIONAL WORK DURING AND AFTER WORLD WAR II

Archaeology in the Chaquaqua Plateau halted during the second world war. Paul Gebhard, who had worked with Renaud in 1934 and 1935 (Cassells 1997), excavated some sites along the Purgatoire River in the 1940s. Results of one excavation followed Renaud's practice of excavating at the base of BGS features. However, Gebhard followed Bartlett's lead by presenting detailed measurements and descriptions of the grinding elements. One rockshelter in Otero County contained rock art, ground groove elements, and boulders with ground stone "depressions" (Gebhard 1943). The excavation yielded projectile points, drills, cores, and other stone artifacts, including a pipe. According to Gebhard (1943), the boulders had "flat pecked tops and evidently served as metates and or anvils. All were deeply embedded in the sterile floor. One boulder had shallow oval depressions circa 180 by 80 mm ground and pecked in one sloping surface. Due to its situation on a sloping surface, this may be a depression in which axes were ground . . . also two pairs of narrow, parallel grooves in which bone awls were presumably ground." Although he found corn in the rockshelter, Gebhard assumed it was intrusive. One of the few to break from the tradition of assuming BGS features signaled developing agriculture in the region.

THE POST-WAR YEARS

The descriptive trend continued when archaeological surveys recommenced in the region in the 1950s. The dominant research agenda continued to search for evidence of the earliest human settlement in North America. Additionally, researchers were interested in building a regional context for the stone enclosures. To that end, surveys and excavations took place along the paths E. B. Renaud, now retired, had taken before the war. Unfortunately, none of the work expanded the understanding of the regional bedrock ground stone features, even though researchers worked in localities where they were abundant.

An example of the general lack of interest in BGS features is the Trinchera Cave excavation. Although the researchers photographed a boulder with bedrock ground stone surfaces during the 1949 description of the site (Figure 1.3), no one attempted to analyze it or neighboring BGS features at the site. It is tempting to believe that the features are so ubiquitous in the region that they are, as Bartlett suggested, somewhat invisible. We return to Trinchera BGS features in chapter 6.

BGS features seem to disappear from Chaquaqua Plateau archaeology until the late 1960s when Robert G. Campbell surveyed the area in his quest to find answers to the broader cultural influences in the region. His research problem centered on whether or not the people living in the canyon systems of the Chaquaqua Plateau were a variant of cultures from the Texas Panhandle, other areas of the Plains, or the Southwest culture area. Campbell surveyed major areas along the Chacuaco Canyon (only one of many on the plateau), looking primarily at the stone circle or stone enclosure sites. During his expedition in the canyons, he located twenty-two sites with BGS features. Unfortunately, his dissertation only describes one of the sites, LA221, in any detail. Nicknamed "metate rockshelter," it is not the "Metate Cave" site (5LA6575) examined in chapter 5. Campbell considered the BGS features indicators of shifts to local agriculture or incipient horticulture as with earlier trends. However, he only paid superficial attention to the features as integral to the archaeological material culture.

PROCESSUAL ARCHAEOLOGY AND CULTURAL RESOURCE MANAGEMENT

In the early 1980s, the development of a US Army training area (Pinon Canyon) took over a large portion of the Chaquaqua Plateau on the northwest side of the Purgatoire River which drains much of the plateau into the Arkansas River. Other major cultural resource management (CRM) surveys during the 1980s include the Comanche Grasslands, the Trinidad State Reservoir, John Martin Reservoir (Eighmy 1984), and Carizzo Ranch (Kingsbury and Nowak 1980). During this period, archaeologists in the region began recording bedrock ground stone features as they found them during CRM or salvage archaeology surveys. Archaeologists made genuine attempts to incorporate the features into site interpretations, mainly as tools used to process locally harvested food resources. The shifted focus added depth to archaeological analysis in CRM reports of the time. Federal law mandated that all cultural features be managed, including BGS features. Tribal consultants evaluated sites with BGS features (Stoffle 1984). Perhaps the most significant change was developing new recording methods on the Pinon Canyon Maneuver Site (PCMS) and Carizzo Ranch archaeology, which captured BGS feature variation. The methods included drawing feature maps to scale, measuring the individual ground stone basins (element) on each feature, taking photographs, and developing a means to specify grinding element measurements. When archaeologists drew feature planviews (scaled drawings) in the field, individual ground stone basins were treated much like rock art motifs (called elements). They assigned each ground stone basin a field specimen (F.S.) number or an element letter (a through z) to re-identify ground stone basins (specifically the measurements taken on ground stone basins) for additional study. During the early stages of this methodological evolution, field documentation changed rapidly. Although it was not always possible to tie specific ground stone basins with their measurements, data are still helpful for broad generalizations about the morphological variation of ground stone basins.

CRM methods evolved under the processual flag. The objective became measurement, evaluation, and building reliable frames of statistical reference to discuss the behavioral patterns inferred from the material remains. In terms of BGS features, the implications for data collection were immense. First, the BGS features had to be described as completely as possible. In effect, they had to be made visible and quantifiable. Second, the features had to be positioned soundly in context with other cultural materials. The latter is much more challenging, especially without rethinking existing assumptions. Thus, advances in the study of regional BGS features quickly stagnated. However, the change in recording methods led to large amounts of data collected and stored with the Colorado Office of Archaeology and Historic Preservation (OAHP).

Descriptive processes under the CRM work necessitated scale drawings of the BGS feature with narrative descriptions of the context. Problems developing feature and artifact classification soon became evident. If each BGS feature had essentially operated as a workspace, every ground stone basin would represent a sub-set of activities at any given feature (Lynch 2017a, b). However, what human behavior did the field record narratives and drawings describe? Are the features a workspace (thus some form of group processing area)? Are the activities at the feature being carried out over days, months, or years? In other words, how much time does each basin represent? How much time does a feature with multiple basins represent? Answering these questions could illustrate multiple grinding episodes over seasonal periods or intensive use over shorter habitation periods. Of course, the types of resources processed might have varied or remained constant. The question became, how to best establish processes that would help solve these problems, at the field-level of observation, with implications at the level of analysis and interpretation.

Archaeologists partially addressed the field observation problem by creating those detailed feature planviews (maps). Each one functioned as visualization tools similar to rock art panel drawings. They sketched individual ground stone basins in site maps; sometimes, they gave them F.S. numbers (or element letters) to include them in site inventories. Increasingly, they patiently measured entire boulders and the individual basins or work area on the BGS features. Attempts were made, in the field, to determine the directionality of grinding motion by looking for striations or pitting within respective grinding basins. Cultural management reports talked about the nearest local resources and other features, speculating slightly about the nature of use-wear and attempting to incorporate the features into site interpretation. By and large, though, the final assessment remained "women ground vegetal materials there" or "people ground food there."

CURRENT INVESTIGATIONS

The search for patterns across the regional canyon system pointed out some problems with the new recording processes as well as some analytical bias. Although regional surveys often included specific metric descriptions (length, width, and depth), these data were scattered across, sometimes large, disconnected paper files. The non-standardized recording processes captured some data, but researchers still had to go to the individual site files to pull metric data for analysis. Without a centralized database for BGS feature data collection or standardized field feature recording methods, the analytical process was cumbersome. The author soon discovered that although regional archaeologists were recording more BGS features, they remained relatively invisible to broader archaeological research.

Cultural resource management forms, accessible from OAHP, reveal another problem with the underlying recording process. The data (length, width, and depth measurements) were largely unreplicable since many feature drawings did not include enough information to match individual basin measurements to the real-world feature. Further, we cannot bring BGS features into a lab for microscopic analysis of use-wear, nor do we have field experiments to illustrate use-wear or natural erosional processes on the local Dakota Sandstone. Thus, the field observer's assessments for striations and pitting could prove unreliable since many features are in open-air contexts. Exposure to natural processes such as wind, freeze-thaw, or water erosion, quickly affects the surfaces of local sandstones. Finally, when tested, observer bias in determining the size and shapes of ground stone basins was extremely high, even under reasonably controlled circumstances (Lynch et al. 2016). Researchers must highly scrutinize historically collected data, especially when discussing the morphological variation in regional descriptions.

Once the analytical gaze focused on BGS features, it became evident that natural processes, in addition to human agency, operated on the Dakota Sandstone. The difference is not easy to determine without specialized experience. In one study (Lynch et al. 2017), the researchers excluded any BGS surfaces found outside of rockshelters because they were subject to more weathering processes. Archaeologists had to scrutinize metric data sets and relocate the features to verify field measurements and use-wear assessments. As of 2020, there were more than 460 sites in southeastern Colorado (OAHP database) described as having bedrock ground stone features. This number probably represents less than 25 percent of the actual number of BGS features in the region. Further, the existing data are highly biased towards the intensively managed cultural resources of the Pinon Canyon Maneuver Site.

Technological developments, especially in photogrammetry (further explained in chapter 5), provide an opportunity to change the way we record BGS. The next adjustment in field recording is substantial because accurate field assessment of BGS features requires time in the field that may or may not be available depending on the nature of the CRM survey. The author's work in a single side-canyon of the Chaquaqua Plateau illustrates this situation. The landowner's permission allowed us to spend at least seven days moving back and forth between sites with features. The time was invaluable. After we located BGS features, we recorded the site using standard, manual field recording techniques (Lynch 2017a, b). We defined each BGS feature, usually a boulder, along with any subset features (workspaces on the boulder's surface) and the individual elements (basins within sub-features or features on a boulder). We recorded every element on the feature planview (map) to scale and in context with the surrounding space. We always used at least two others from the team to verify the recorded data. Once drawn and photographed, we removed all element markers. We revisited the site at least twice within the next couple of days, at different times of the day because the lighting changes the perspective of the sandstone. A team member other than the original recorder checked the maps. The second team member resolved replication issues. This example outlines the ideal recording situation for BGS features. It may be unusual to have the time and number of trained team members to perform the manual recording process but quickly sketched field maps combined with 3D recording provide quick, reliable results using standard field equipment.

Close range photogrammetry (CRP) is the best method to ensure accurate metric data are the foundation of comparative shape analysis. Photogrammetry is becoming more common for archaeological data collection and analysis. However, it is still in the nascent stages of application and publication. The methodology for field collections will probably change a little over the next few decades. The lab processing and analysis will alter a great deal as more individuals explore the boundaries of the technology.

Nonetheless, we should continue to record sites using pen and paper, especially as part of engaged teaching pedagogy. The most important reason for teaching manual feature recording is developing critical spatial analysis skills in future archaeologists. As the field moves towards virtual reality, this aspect of the learning process will continue to shape how we think about BGS features.

Close range photogrammetry is the process of taking stereo-overlapping photographs in the field for processing in a lab where archaeologists reconstruct a virtual image of the object using modeling software. The final product is a 3D representation of the real-world object with many of the same real-world properties, including texture, 3D context, shape, and size. The model can be measured, manipulated, and experienced using other software programs with a V.R. headset or CAV technology. Since BGS features are generally difficult to reach and because of the high interobserver error in field recording, it is possible to create scientific collaboration about these features using virtual reality technology. Because we are still learning to understand BGS features and our bias, it is essential to include as many voices as possible. These virtual workspaces include scientists, students, interested private individuals, but especially descendant communities.

Bedrock ground stone features add to our knowledge of prehistoric land use and social worlds. However, before we can adequately understand how and where they fit into the existing archaeological context (primarily projectile point technology, pottery, and architecture), we must identify the material objects then define them in space and time. We have to standardize the methods used to describe the BGS features and the human agency at those features. Using terms like "metate," "mortar," "basin," "slick" is fine, but they have fuzzy implications such as food use, agriculture, or perhaps they anthropomorphize natural phenomena. The remainder of the book will lay out a scientific process for describing, analyzing, and interpreting those features already documented by researchers in southeastern Colorado. We are still developing the methods and processes. However, this book represents the opportunity to create a space to think about the nature of these spaces in context with neighboring archaeological features and artifacts.

At the heart of the scientific method is transparency. One's data must be available for cross-examination and refutation. Because regional BGS features have been invisible and neglected for so long, we should envision them as crucial evidence of human cultural behavior worthy of study. The features are real, so they must be examined, especially in light of the surrounding archaeological context. 3D visualization is one way to expand our ability to further engage with the role these features played in prehistory. The rest of this book will look at individual examples that illustrate this vision.

NOTE

1. Janet Chatin's field notes and papers are curated at the Huerfano Heritage Center in Walsenburg, Colorado.

2

Comparative Analysis of BGS Features and Terminology

Bedrock ground stone features found on the Chaquaqua Plateau create a puzzle for archaeologists interested in how past societies may have inhabited regional landscapes. As a distantly orphaned material culture in the region, one avenue of inference is to compare the Colorado features to those found elsewhere with descendant voices to speak for them. While BGS features occur across North America, those geographically closest to southeastern Colorado, with ethnographically documented functions occur in California and the Great Basin regions. The widespread distribution of bedrock ground stone (BGS) features across North America suggests these archaeological phenomena were well-entrenched cultural traditions (Bennyhoff 1953, 1956; Claassen 2011; Dill 1975: Driver and Massey 1957; Lynch 2017a, b). What is not apparent for many regions, without direct ethnographic associations, is how and why some cultures relied on BGS features for food processing or other types of raw material manufacture (Lynch 2014; Lynch 2017a, b). Although research is beginning to focus on these features in other areas of North America (Claassen 2016), published measurement data connected to human communities are rare.

To build a descriptive framework for BGS features in southeastern Colorado that tests current assumptions while also providing valuable information to answering the how and why questions, this chapter compares the morphology of element shapes (grinding surfaces) from the High Plains to those from two areas in California. There is no ethnographic information for the High Plains, whereas consultants from descendant communities have confirmed the cultural behavior associated with the observed archaeological signature in California (Figure 2.1). An underlying assumption of this inference process is that similar shapes indicate similar functionality. In other words, processing acorns, pinon, or other food resources leaves behind a distinctive pattern, which may not be the case. However, a fundamental step in building a reference system for the Chaquaqua Plateau region is determining what the observed signatures could mean to landscape use or human behavior. Reasoning from analogy, although not absolute, provides baseline information that helps test old inferences while building new grounds for future analyses.

A lively subject in the study of BGS features is the interpretation that they are part of socialized landscapes, in which people worked together to process food or other resources. Dick-Bissonette (1998) and Jackson (1991) made compelling arguments for the gendered use of acorn grinding features in California with consequences to land use and tenure. Lynch (2017a, b) and Lynch and colleagues (2017), inspired by Claassen's (2011) research with bedrock features in the southeastern United States, have argued that BGS sites in the JE Canyon area offer evidence for seasonal resource processing by related groups of people over an extended period of time; from hundreds to perhaps as long as a thousand years. Lynch and colleagues (2017) illustrate the possibility of group work, thus providing support for the idea that the BGS sites were locations where the social reproduction of knowledge could have occurred through ritualized landscape use. Basso (1988) provides convincing descriptions of how humans map social behavior onto their lived environments. Since the grinding activities are highly socialized events (see chapter 8), the spaces in which they occur affect how people think about their broader environment and how they move through the landscape.

Much of the BGS feature research in southeastern Colorado relies heavily on ethnographic evidence from other grinding cultures, particularly those in California, to provide frameworks for interpreting the patterns observed in the local archaeological record. At the heart of ground stone research in southeastern Colorado is the assumption that most, if not all, of the BGS features, represent food-processing (Campbell 1969; Gebhard 1943; Loendorf 2008, Loendorf and Loendorf 1999; Lynch 2017a, b; Owens et al. 2000, 2007; Renaud 1930; Renaud and Chatin 1943; Schiavitti et al. 2001; Stoffle 1984). A corollary assumption asserts that women were the primary producers and users of BGS features (Loendorf 2008, Loendorf and Loendorf 1999; Lynch 2017a, b; Owens et al. 2000; Owens and Loendorf 2004). Support for these assumptions derives from the well-documented California acorn-processing traditions (Barrett and Gifford 1933; Gayton 1948a, 1948b; Gifford 1932, 1936, 1967; McCarthy et al. 1985; True 1993). To explore whether there is enough physical similarity between the regional BGS features to warrant the continued reliance on this model, this chapter compares the shape of ground stone elements in California with those found on the Chaquaqua Plateau.

The metric shape descriptions, developed from published data sets of BGS features documented on JE Canyon Ranch (Lynch 2017a, b), are compared to three studies of well-defined food processing contexts in California (McCarthy et al. 1985; True et al. 1991; True and Waugh 1981). Since the goal is to understand possible behavior directly related to the material culture, the comparative data set is limited to those sites with Native American descendant community's consultation. Lynch (2017a, b) defines the methods of shape analysis for BGS features. The methods define comparative shapes from a functional study of bedrock mortars supported by ethnographic data from Mono peoples in California (McCarthy et al. 1985), a description of types of bedrock mortars, metates, and slicks related to ancestral populations of the modern Luiseño (True et al. 1991), and a descriptive study of prehistoric bedrock features, also related to the modern Luiseño people, along Frey Creek, San Diego County, California (True and Waugh 1981).

The underlying assumption for this type of comparative analysis is that similar processes should create similar ground structure tool kits. We should expect that artifacts grinding identified as "metate," "mortar," or "slick" by ethnographic sources California would resemble BGS shapes observed in southeastern Colorado also called "metate," "mortar," or "slick." Groups of people (related by cultural, linguistic, and social ties inhabited the canyon along Chacuaco Creek) should do similar things (for example, food processing). Ground stone tool patterns should emerge resembling shapes created by similar processes in areas with good ethnographic context. The comparison provides a structure to discuss patterns or trends when considering how groups and individuals once occupied BGS spaces.

Directly comparing shapes between regions affects how we think about regional subsistence strategies through time. There should be a difference between mobile hunter-gatherer strategies (southeastern Colorado sample) and foraging strategies such as those focused on acorn or pinon crops (California sample). In addition to better understanding subsistence strategies, significant differences in BGS morphology could point to different cultural grinding practices between the regions (the western United States versus the High Plains). Comparative research supports statements about how BGS spaces in Chacuaco Canyon could have been used by household units or extended families harvesting and processing reliable local foods, as is the case in California, or whether they represent a unique grinding process distinctive to High Plains cultures.

ETHNOGRAPHIC CONTEXTS IN CALIFORNIA

Ethnographers documented wild food processing at bedrock features (Barrett and Gifford 1933:142; Gayton 1948a, 1948b; Gifford 1932:23; 1936; Kroeber 1976:88, 293, 411, 649; McCarthy et al. 1985). Archaeological analyses of these features explore settlement patterns and the functional side of morphological variation (Bennyhoff 1956; Dill 1975; Fenenga 1952; McCarthy et al. 1985; Parkman 1994; Payen and Boloyan 1963; True 1993; True et al. 1991; True et al. 1974). Recent archaeological studies explore large scale human behavioral analyses (Breithopt, 2010); however, bedrock mortar sites have been used to understand the settlement size reflected by the number of mortars on features found at specific localities (Table 2.1; Bennyhoff 1956; Parkman 1994; True 1993; True et al. 1991). The data in this chapter come from two Native American communities, the Mono and the Luiseño.

Mono Cultural Contexts

Mono peoples were described based on their territory east or west of the Sierra Nevada (Gayton 1948a, 1948b; Gifford 1932; Kroeber 1976). Gifford (1932: 16) reported specific linguistic terms to identify those living on either side of the Sierra Nevada, which might be evidence for the cultural distance between the two groups. The Western Mono oriented towards the San Joaquin River, rather than the Great Basin, and south of the San Joaquin watershed (Gayton 1948b: 212, 256, 257; Gifford 1932: Fig. 1). However, as with many landscape divisions, this one is a heuristic device. The Western Mono maintained regular cultural and trading ties with the Eastern Mono (Gayton 1948b: 214, 258–259). They influenced (and were influenced by) their nearest neighbors, the Yokuts to the west and Miwok on the north (Gifford 1932: 15).

Western Mono settlement patterns cluster around the springs and slopes of local drainages. Gifford (1932: 17) reported that these landscape features were named for, or in relation to, nearby landmarks. Kroeber (1976: 586) wrote, there are "six named western Mono divisions, one each, roughly speaking, on each side of the three great streams that flow through their territory." Gifford (1932: 17) divided their "living sites" into camps and villages with three main components: proximity to water, relative slope, and presence of outcropping granite boulders for bedrock mortars. Western Mono traveled across their landscape on regular seasonal rounds. Groups moved from lower elevations in the winter to higher ones in the summer and after deaths (Gifford 1932: 17). Gayton (1948: 213) states that groups always followed the same routes.

People tended to regularly visit the same locations to collect and process seasonally gathered foods. Gifford (1932: 17) observed, among the Western Mono, an "absence of ideas of private or family ownership of acorn places, fishing places or other food gathering places." However, Gayton (1948: 214), in discussing land tenure, argued: "certainly there were no rigidly defined limits, but the fact that a bride returned to her mother's gathering place rather than going to her mother-in-law's or better yet gathering where she would, indicated [that] at least a tendency toward personal association with a gathering place, if not indisputable proprietary rights."

While Western Mono lacked centrally located, large villages (Gifford 1932:19), they did have smaller "hamlets" that were seasonally occupied (Gifford 1932: 19). Habitation sizes "ranged from 1 to 8 huts (average 3)" that could accommodate between "1 to 39 (average 13)" individuals depending on whether it was a "camp" or a "hamlet" (Gifford 1932: 19). People related by blood or socio-political alliances occupied the same habitation areas (Gifford 1932: 19) and consisted of habitation structures, acorn granaries, sweat houses, and rock outcrops for bedrock mortars (Gifford 1932: 20). Gayton (1948b: 213) noted the differences between camping sites (some specialized for fish and hunting, others for seed gathering); but also characterized "large villages" as those with around "25 houses."

Luiseño Cultural Context

The landscape of the Luiseño, historically hunter-gatherers, is in modern San Diego County, California (Sparkman 1908). Sparkman (1908:190) described the organization of their territory as, "each band had its track in the San Luis Rey Valley, and another on Palomar, to which it moved during the acorn-gathering season. The land sometimes again sub-divided among different families of which the band was composed." Unlike the Western Mono, they protected their territories with the "greatest jealousy" (Sparkman 1908: 190). Names of places were primarily of plants and associated animals (Sparkman 1908: 190). The Luiseño harvested various food and seeds, but acorn was the staple (Sparkman 1908: 194). Sparkman (1908:207) described the bedrock outcroppings as "ordinary mortar holes in a large flat granite rock near the dwelling."

The Luiseño may have migrated to the area between 2,000 and 3,000 years ago (White 1963: 91). White (1963: 99) divided their territory into 50 rancherias of about 200 persons. In their past, women gathered plant resources, but as with the Western Mono (see Gayton 1948a, 1948b), "both sexes without regard to age turned out *en masse* to harvest the acorn crop which was carried with trumpline and net to large granaries in the villages" (White 1963: 116). White (1963:116) suggests the "masse" harvest was due partly to the short harvest season, the reliance on the acorn crop, and human competition with birds and rodents.

Although characterized as hunter-gatherers, White (1963: 116) calls the Luiseño a "relatively sedentary population being in one place" because of the rancheria system, which acted "in some ways similar to a town." However, as described by Sparkman, the rancheria system was divided into self-sufficient landscapes "within its boundaries, for example oak grove, chaparral-covered slopes, river bottoms, springs and so forth" all accessible within "half a day" (White 1963: 116). There were "semi-permanent villages" with patrilineal and "some matrilineal social controls" (White 1963: 120; 172).

White (1963:123, 124) describes different ownership classes, including collectively owned areas, used for more than food resources. Women used family gardens but "inherited patrilineally and used patrilocally" (White 1963: 123, 124). The kinship category applied to the "sharing of certain fixed facilities such as bedrock mortars" (White 1963:125). White (1963:125) describes "a number implements worked into the nearby outcropping boulders." He describes smooth bedrock metates, bedrock mortars with shallow side elements (used for the temporary storage of seeds, acorns, or condiments during food preparation), 'portable' mortar, shallow bedrock basins (used to dry acorns and seeds before grinding) covered with a basket when in use, large basins with polished surfaces (perhaps for soaking and dressing skins, reeds and other basketry materials), and pitted boulders for sharpening wood and bone points. Bedrock feature ownership varied. Some, the hide-processing features or tool-sharpening features, could be used by anyone; but the bedrock mortars belonged to the family group (White 1963:125).

ARCHAEOLOGICAL INFERENCE AND BEDROCK MORTARS

Working in the Yosemite region, Bennyhoff (1953, 1956) noted clustered mortar features at elevations ranging from 3,999 to 10,699 feet above modern sea level. House sites and village sites (small and large) were rarely above 7,999 feet in the Yosemite areas, while the number of small and large campsites increased at higher elevations. Bennyhoff (1956:13) hypothesized that people selected site location because of critical resources, including water. Both acorns and pine nuts were brought to the villages for storage and processing from the harvest location (Bennyhoff 1956:13).

Bennyhoff (1956:13) suggested that the mortar depth reflected the length of local occupation and intensity of mortar use (a house site averaged 7.62 cm or less, a small village site averaged 15.24 cm or deeper). The depth could also reflect available milling space on the bedrock surface—inhabitants were less likely to abandon a mortar due to an awkward or inefficient depth if there were no readily available space for new mortars on the boulder outcrop (Bennyhoff 1956:13). Bennyhoff (1956:13) reflected on the possibility that "each woman had some claim to a specific mortar hole or holes and the amount of use might vary with the individual" and hypothesized that the number of mortars would predict the size of the community associated with the processing center (Table 2.1). Smaller family units were associated with smaller clusters of mortars, while larger villages were associated with more than 20 mortar clusters.

Parkman (1994:46), building on the Yosemite model proposed by Bennyhoff, advocated a similar link between the size of pre-colonial Ohlone settlements and the number of mortars found at archaeological sites in the San Francisco Bay area. Parkman (1946:46) extended the inference to include the social organization associated with the relevant occupation patterns. Fenenga (1952) excavated a small village settlement in central California. He determined it was a community of about thirteen households based on the midden and house pit features. The small village

Table 2.1. Summary of Archaeological Inferences for Bedrock Mortar Sites

Number of Mortars	*Size of Community*	*Locale*	*Source*
1–7	House site	Yosemite	Bennyhoff 1956:12–13
8–19	Small village		
21+	Large village		
1–5	Single-family	Northern California	Parkman 1994:46
6–10	Single extended family/clan		
11–19	Clan/Moiety		
20 +	Moiety/village		
11 mortars/17 metate-slick	Marginal village/Short Term use	SDi-5589	True 1993:3–5, 8
	Large habitation area	SDi-7210A	
8 mortars/59 metate-slick	Summer Camp/Village/Acorn Place	SDi-539	
		SDi-682	
46 mortars/5 metate-slick	Major village (TOMKAV)	SDi-308	
	Major village (MOLPA) "major occupation"	*(San Diego County)*	
114 mortars/81 metate-slick			
286 mortars/109 metate-slick			
21–48 mortars, shallow pits, deeper pits	Small village community (13 households: *Yokuts/Mono*)	Slick Rock Village	Fenenga 1952:342

had between 21 and 48 bedrock features (Table 2.1), thus supporting the archaeological inference about the relationship between settlement patterns (size) and the presence of bedrock features noted in other areas of California.

Patterns in bedrock ground stone surfaces described by True and colleagues (1974; 1991) and True and Waugh (1981) differed from those observed by Bennyhoff, and to some extent, those described by Parkman. However, the association of bedrock mortar density with the size and intensity of site occupation remained germane for archaeologists in other areas of North America, particularly those in southeastern Colorado, so it is worth noting the historical thought process. The following comparative approach is designed with these archaeological assertions in mind. It seeks to test the notion that the forms we observe and call *metate, mortar, slick* in Colorado resemble those discussed in California because these names have functional, demographic, and social implications.

DATA SETS

The data in this comparison were published by McCarthy and colleagues (1985), True and Waugh (1981), and True and colleagues (1991) for California sites, and by Lynch (2017a, 2017b) for Colorado BGS sites. McCarthy and colleagues (1985) condensed measurement data from three different sites located across Western Mono traditional territory (Bass Lake, Manzanita Lake, and Kerckhoff Reservoir). They conducted a functional analysis of mortars in conjunction with descendant community consultation. They categorized three primary types: starter mortars, finisher mortars, and seed mortars. Local Mono consultants, who indicated that depth of the milling surfaces was a primary means of identifying the production process and resources processed, added a fourth functional type of milling surface, slicks, used to crush manzanita (McCarthy et al. 1985: 324–327). Thus, consultation tied typology (based on shape) to specific food resources but also production processes. Starter and finisher mortars were used to process acorns into fine flour. Seed mortars were used to process locally harvested seeds. Slicks were used for berry processing (Table 2.2).

True and his colleagues, working in Southern California, distinguished between five types of bedrock ground stone surfaces: mortars, metates, slicks, pocket collars, and cupules. They defined mortars as ground surfaces with a round planar morphology between 10 cm and 30 cm in diameter, with a depth between 2 cm and 30 cm and a u-shape cross-section (True et al. 1991:63–74). They used the term "metate" for grinding surfaces with a round to oval shape, definable margins, and an average

Table 2.2. Summary of Bedrock Mortar Functional Data in California

Site Name or Mortar Description	*Mortars(N)*	*Range: Diameter (cm)*	*Range: Depth (cm)*	*Function*	*Ethnographic Affiliation*	*Source*
Slick Rock Village	94	10–25	3–23	acorn seeds	Sierra Nevada: Yokuts and Mono	Fenenga 1952: 342
Starter	28	7.3–16.5	.8–6	acorn	Sierra Nevada: Mono	McCarthy et al. 1985: 327*
Finishing	15	15.4–20	5–12	acorn		
Seed	22	15–22.75	8.2–21.5	seeds		
Slick	2	17–19.5	.5–1.4	manzanita		
Ama-14	1,158	7–16.5	1–25	seeds	Northern Miwok	Payen & Boloyan 1963:2–5
Mortars	–	10–30	2–30	–	Southern California:	True et al. 1991:63–64
Metates	–	–	.5–4	–	Dieguero or Luiseno	
Slicks	–	–	0	–		
Mortar Collar	–	–	–	–		
Cupule	–	–	–	–		

*Data are from the consultant model.

depth of .5 cm to 4 cm. They described slicks as ground surfaces that had no definable shape and no discernible border. Other shape categories included the mortar collar or pocket mortar, often associated with a mortar proper, and cupules, often located on vertical aspects of the boulders. Researchers hypothesize that these other shapes had been used to hull seeds or as part of ceremonial events (True et al. 1991; White 1963).

Bedrock ground stone features in southeastern Colorado have yet to be directly associated with descendant communities, although the Comanche, Shoshone, Southern Ute, and Jicarilla Apache are historically associated with the canyon landscape (Flint and Flint 2005; Hammond and Rey 1940, 1966; Stansbury 1988[1852]). Campbell (1969) suggested the BGS features may have been used as early as 2,000 years ago, then abandoned around 500 years ago. The chronology will be unpacked in chapter 7. In the absence of direct ethnographic data, Lynch (2017a, b) and Lynch and her colleagues (2017) relied heavily on the California research to build a working model of BGS use. Describing 206 BGS surface over eleven sites in a side canyon, (Lynch 2017a, b) proposed that the sites were used by a group of closely related individuals over a long period of time, with land tenure like those reliant on acorns from California. However, the range of sizes and depths in the sample suggested that some sites were not used solely to process food resources (Lynch 2017a). Another pattern emerged from this research. Unlike the California models, the sites with dense clusters of BGS surfaces occurred in rockshelters or enclosed spaces. Open-air features were more lightly used or dominated by slick-like ground stone surfaces.

The terms "bedrock mortars" and "bedrock metates" are still widely used when referring to any type of boulder with ovoid-like surfaces ground on them. Lynch (2017a, 2017b) preferred to ignore terms like "metate" or "mortar," relying instead upon geometric and morphological descriptions because of the close association of mortar and metate to specific modes of food processing, as yet untested for the Chaquaqua Plateau. Thus, it became less problematic to refer to all ground stone surfaces on boulders as BGS features.

METHODS

Data collection methods varied slightly for each sample. For instance, McCarthy and colleagues (1985:321–323) collected depth measurements following Dill (1975); while Lynch (2017a) collected data guided primarily by Adams (2002) but further informed by Bartlett (1933), Dill (1975), Eddy (1964), McCarthy and colleagues (1985), Mohr (1954), True (1993), True and Waugh (1981), and Woodbury (1954). The standard data for ground stone surfaces include the major axis (length), minor axis (width), and depth. McCarthy and colleagues (1985) tailored their methods to consider slope, including two measurements of surface diameter.

Lynch (2017b) used five morphological variables to approximate the shape of BGS surfaces in the Chacuaco Canyon sample. Three (circularity, form, and elongation, abbreviated to CFE) illustrate the planar surface shape. Two (perimeter in centimeters and area in centimeters squared, abbreviated to PA) describe the relative surface size (formulae are described in Lynch 2017b). McCarthy and colleagues (1985:330) offered a new descriptive variable, "configuration," which expresses the ratio of mean diameter to depth (MD/d). While Lynch (2017b) did not use this variable, it is recreated in this chapter to estimate a three-dimensional shape for comparative purposes.

The published BGS measurements from McCarthy and colleagues (1985:347–356), True and colleagues (1991:63–69), and True and Waugh (1981:91, 94, 95, 100, 101) were entered into SPSS along with those published by Lynch (2017b). Both True and colleagues and Lynch use length and width measurements; McCarthy and colleagues used diameter. Each group of researchers differed in their collection of depth measurements, so they are used in this study broadly as indicators of shape for comparison only, not to imply specific functions, time, or production volume.

Since BGS terminology may have functional implications, and because so little is known about the function of BGS in southeastern Colorado, Lynch (2017b) coded each surface based on observable shape (for example, ellipse, d-shape, circular). Lynch's (2017b) shape descriptions are compared against McCarthy and colleagues (1985), True and colleagues (1991), and True and Waugh's (1981) functional terminology (e.g., starter or finishing mortar, metates, slicks, etc.), which have been ethnographically described. Thus, it is possible to relate the mathematical description of the shape and the geometric description of surface morphology to compare the surface shape with functional assessments (metate, mortar, slick, etc.) from the California sample.

RESULTS

The Mono cultural sample, from four different localities, has 146 bedrock surfaces. The Luiseño cultural sample is from two different localities with a total of 218 bedrock surfaces. The Chacuaco cultural sample, with 126 grinding surfaces, comes from nine sites in an interconnected canyon system, each within easy walking distance of each other (Table 2.3).

McCarthy and colleagues (1985) found that depth was a good indicator of functional type. Lynch (2017a) relied on depth to measure how much

Table 2.3. BGS Grinding Surfaces in the Regions Sampled

	Sites (N)	*Surfaces (N)*	*Sample (N)*
JE Canyon Survey	11	208	126
Mono culture	4	291	146
Luiseño culture	5	304	218

time was spent grinding at the basin, thus, a time investment. Results for the mean depth, size variables (perimeter, area), shape variables (circularity, form, elongation), and McCarthy and colleagues' (1985) shape variable "configuration" indicate a substantial difference between the regional samples (Table 2.4).

The depth, size, and shape variables do not represent a normal distribution within the samples. Non-parametric tests indicated that the variables between regions sampled are distributed the same (p = .000). In the Mono sample, depth is for sites sampled (Independent Samples Kruskal-Wallace Test p = .088), size variables (PA) are different (Independent Samples Kruskal-Wallace Test p = .023). The shape variable *configuration* is the same across sites (Independent Samples Kruskal-Wallace Test p = .060). Shape variables CFE equals 1.0 because the variable *mean diameter* was used, which means all planar shapes conform to a circle.

In the Luiseño sample, none of the variables are the same across sites (Independent Samples Kruskal-Wallace Test p = .000). Whereas in the Chacuaco sample, depth, size variables (PA), and configuration are not the same across sites (Independent Samples Kruskal-Wallace Test p = .000), but the shape variables CFE are the same (Independent Samples Kruskal-Wallace Test p = .496). The shape variable configuration relies on depth as an approximate z-value which accounts for the discrepancy between results of the shape variables.

Table 2.4. Summary of Shape Variables by Region

	Northern California			Southern California			Southeastern Colorado		
	N	*Mean*	*Std Dev*	*N*	*Mean*	*Std Dev*	*N*	*Mean*	*Std Dev*
Length	146	15.81	3.38	297	20.40	8.12	147	23.19	8.19
Width	146	15.81	3.38	297	17.42	5.17	147	14.58	5.97
Depth	146	7.13	4.82	219	7.40	5.28	146	1.70	1.04
Perimeter	146	49.68	10.60	297	59.69	20.13	137	60.16	21.08
Area	146	205.3	86.4	297	305.4	284.6	137	290.70	197.10
Form	146	1	1	297	1.18	0.47	137	1.76	0.66
Elongation	146	1	1	297	0.89	0.16	137	0.62	0.17
Circularity	146	1	1	297	0.94	0.09	137	0.78	0.11

The depth variable was a proxy for the function of a particular BGS surface (often deeper implies a mortar, or the pounding of a resource) and as an indicator of the amount of time used (Lynch 2017a; Wright 1993). In the southeastern Colorado sample, the depth of bedrock ground stone surfaces ranges from 0 to less than 5 cm (mean = 1.7294), whereas the California samples illustrate a wide range of depths (Figures 2.1, 2.2). The situation was not wholly unexpected because of the multi-step processing of resources by the Western Mono and Luiseño peoples. Depth and size variables overlap between the Mono cultural sample and the Luiseño cultural sample (notably at TomKav), but the Chacuaco sample overlaps with the Luiseño Frey Creek cultural sample.

Size variables (PA), although significantly different between the regions, do not exhibit the skew observed in the depth variables. The one outlier is from the Frey Creek sample in which the bedrock slick has the largest perimeter and size of all the BGS surfaces sampled. Shape variables (CFE) illustrate the difference between pounding resources (often recorded as circular; C = 1.0; F = 1.0; E = 1.0) in California and the use of other processing behaviors: notably push-pull motions or push-pull circular motions. The shape variables illustrate the presence of tool grooves (C = .40; F = 4.0, 6.0; and E = .20) in the Chacuaco cultural sample.

The interchangeability of terms between regional archaeological contexts, for example, mortar, metate, or slick) is awkward because the descriptors often imply functionality. The variables thought to indicate use and function are depth, size, and shape. The Chacuaco sample variables for size overlap more closely with the Luiseño categories of "metate," "pocket," and "slick," as wells as the Mono categories of "starter mortars." In terms of the shape variables CFE, there is no clear distinction. The Chacuaco cultural sample overlaps the relatively shallow shape values and the predominately deep mortars of the Luiseño sites. The Mono sample is not shown because the value is 1.0 across CFE, indicating a circle. Rounded deep basins like these are rare in southeastern Colorado. McCarthy and colleagues' (1985) configuration variable illustrates a clear overlap between the "elliptical" shape sampled in Chacuaco Canyon, the Luiseño "metate," and Mono "starter mortar."

DISCUSSION

The comparison of three different cultural areas, all reliant on bedrock outcrops for processing staple resources, illustrates the variation between regions. The Western Mono sample is not complete since it includes only four large sites from a dispersed area, whereas the Luiseño and Chacuaco prehistoric samples are from interconnecting landscapes. One objective was to compare the relationship between functional, shape descriptions of BGS surfaces between ethnographically supported regions with the Chacuaco cultural sample as accurately as possible. Another objective was to develop an inferential foundation for describing BGS feature sites in southeastern Colorado by comparing the regional sample with an ethnographically controlled sample. The results do not solve the functional problem for the Chacuaco Plateau.

Nevertheless, the results support four conclusions: 1) the BGS surfaces most closely resemble (in terms of size and shape) the functional surfaces called "metates" by the Luiseño or "starter mortars" by the Western Mono; 2) the depth differences between the regions supports the idea that push-pull and push-pull-rotary grinding motions are most likely for Chacuaco side canyon; 3) functional use, though not as diverse as the Luiseño peoples, was probably reasonably diverse, meaning we should expect that not all surfaces were used for food processing; and 4) the differences between depth, size, and shape characteristics across regions dictates that inferences about population size and social structure using California archaeological models, must be scaled down for the Chacuaco side canyon.

The resemblance of Chacuaco prehistoric bedrock ground stone surfaces to "starter mortars" or "metates" is based on the planar surface, size, and shape variables. The results do not support the use of the functional term "mortar" for the Colorado area. The circular shapes in the Chacuaco sample are small, conical shapes whose function remains unknown. The depth of the surfaces coupled with the lack of pestles suggests rotary grinding more than pounding behaviors indicated by the functional term "mortar." Arguably, pestles may have been used periodically at these "mortar-like" surfaces. If so, their absence could be explained by using wood tools, which would rapidly degrade, or by stone pestles that were curated or repurposed, thus are no longer in association with their proper toolkit.

The application of the term "metate" in references to regional Colorado bedrock ground stone surfaces (for instance, "bedrock metates" or "BMETS") should be discouraged because we do not yet know which of these surfaces are related to food processing and which were used for tool making. An example of possible confusion would be at site 5LA11451, in the JE Canyon sample. The site has significant deviations in size and shape variables that led Lynch (2017a) to conclude that some surfaces were used to process foods, while others were probably used for tool making. Applying the term "metate" to all surfaces leads to erroneous interpretations of how 5LA11451 was used by people inhabiting the site, whereas the generic term BGS surfaces exert no such influence. Additionally, the terms "mortar" and "metate" can be viewed as colonizing terminology that minimizes original cultural meaning associated with these types of spaces, thus further usurping descendant community control of their natural and cultural past.

The depth differences between the California and Colorado samples dictate a cautious approach when attaching labels from other regions to an understudied phenomenon. Lynch (2017a) used the depth variable to describe the amount of time investment in sites along the Chacuaco side canyon; she used Wright's (1993) use-wear formula based on grinding push-pull fashion. The parent material was Dakota Sandstone in each instance, but Wright used a two-handed mano, for which there is no evidence in the Chacuaco artifact assemblage. The question remains whether depth should be used as a functional descriptor to indicate time investment or whether it should be used at all. The same question haunts McCarthy and colleagues' (1985) discussion. The present study does not help to clarify the situation but instead provides evidence that caution should be used before using depth to infer functional types or site inferences.

The Chacuaco prehistoric BGS sample is different from the California sample to the extent that direct inferences to social structure or population structure should be made thoughtfully, if at all. However, we should construct some inferences based on the evidence discovered at the fourteen sites within the Colorado side canyon's bounded landscape (Table 2.5). Three of the sites (5LA12783, 5LA12784, 5LA12785) are small areas with tools and lithic debitage, which could be the remains of campsites related to the eleven BGS feature sites. 5LA11455 is near an open terrace with lithic debitage, which could represent a small campsite. Finally, the presence of tools and lithic debitage at the head of the drainage system suggests another possible campsite. Although there are no direct links between these "possible camps" and the BGS sites, the landscape is relatively easy to traverse, taking no more than a few hours to walk between the most distant of the localities. Furthermore, there are no insurmountable landscape features between any of the sites in the canyon, so that they can be considered provisionally related.

The archaeological inferences from California offer some support. However, it should be used carefully for developing inferences about the population and social structure of the people who inhabited the Colorado side-canyon. Since the diversity of BGS surfaces most closely resembles the Luiseño culture, an argument can be made that the archaeology in the Chacuaco samples represents a group of people related by family connection with clan and possibly moiety level social ties (Table 2.5). Site 5LA11453 with limited rock art could have been used by a more extensive aggregation of people (based solely on the number of BGS elements, not the presence of other archaeological features). If so, the people were mostly related family members but may have been connected at the clan

Table 2.5. Summary of Chacuaco Canyon Sites with Interpretations

Site number	*BGS elements (N)*	*Possible population size supported*	*Social structure*	*Rock art*	*Interpretation(s)*
5LA11453	68	aggregation of people, related	extended family; moiety	over pecked on four BGS surfaces	Seasonal food processing
5LA11454	10	single extended household	extended family; clan	large boulder; pecked abstract, anthropomorphs, zoomorphs	Ritual related: possible first fruits ceremonial
5LA11450	1	individual	extended family; clan	large boulder: pecked abstract, anthropomorphs, zoomorphs	Ritual related
5LA11455	24	small family or household	family; clan	2 small panels: abstract; small zoomorphs	Seasonal food processing or possible puberty or food rituals
5LA12781	4	small family or household	family; clan	none	Seasonal food processing or possible puberty or food rituals
5LA12782	7	small family or household	family; clan	none	Seasonal food processing or possible puberty or food rituals
5LA11451	32	extended family or village;	extended family; clan	three panels: pecked abstract; zoomorphs; pecked cupules; grooves	Possible habitation; seasonal food processing or possible puberty or food rituals
5LA12788	2	small family or household	family	none	Seasonal food processing
5LA12787	8	small family or household	family	none	Seasonal food processing
5LA11446	29	extended family or village;	extended family; clan	large panel: pecked quadrupeds	Ritual related: possible first fruits ceremonial
5LA12786	7	small family or household	family	none	Seasonal food processing
5LA12783	0	unknown	unknown	none	Tool and lithic artifacts related to camp
5LA12784	0	unknown	unknown	none	Tool and lithic artifacts related to camp
5LA12785	0	unknown	unknown	none	Tool and lithic artifacts related to camp

level. It is the only site large enough (in terms of quantity of BGS elements and boulder feature size) to accommodate many workers and related family members.

Small family units, with possible extended family or clan level associations, could have been associated with 5LA11454, 5LA11455, 5LA11446, and 5LA11451. The rock art at these sites suggests a ritual purpose to the material processing. Further, 5LA11451 consists of multiple rockshelters with a large midden. Loendorf (2008) suggests the site could have been a long-term habitation area, but it could also have been used for various ceremonial purposes (see Claassen 2011). Five sites (5LA12781, 5LA12782, 5LA12786, 5LA12787, 5LA12788) could have been used by smaller family-related groups or extended family groups clan level. The BGS feature at one site, 5LA11450, was probably not used to process food resources. The surface is relatively flat and associated with a large rock art panel, nicknamed "Archaic Maze," and a second hunting panel in the adjacent rockshelter (Loendorf 2008). Although we do not know with certainty, it is probably safer to assume that the surface was more closely related to the rock art panels, perhaps tool sharpening, than to food processing.

If larger groups of people used the Colorado side canyon, we should expect, based on the California ethnography, evidence of storage facilities near the grinding features. The problem is the recognition of storage facilities in the archaeological context. For instance, were resources buried or stored in facilities made of skins, wood, or other quickly vanishing materials? There are potentially three archaeological stone features that could be the remnants of ancient storage. However, these features have not yet been excavated.

The small Colorado canyon was probably inhabited by a culturally related group, who considered the side canyon landscape theirs, or at the very least the fruits of the side canyon as their "breadbasket." However, sharing access resources (including the availability of ground stone workspace) between an extended social network should be expected. There may have been some ritual attached to hunting, seasonal fruit harvesting, and or rites of passages as "special" sites marked by the rock art if the BGS and rock art panels are contemporaneous. Finally, while most of the sites were probably used by small, related family groups, at least one could have been used by a larger, perhaps clan to moiety level group.

CONCLUSION

The comparison between California and Colorado regions offered some insights into the archaeology of BGS features on the High Plains. Perhaps the most important lesson is that through the description of these features, we will eventually understand how they fit into regional models of landscape use. Although we must use caution when applying ethnographic models outside the region, they represent a rich resource to start constructing regional inferential frameworks.

The landscapes of the Western Mono and Luiseños resemble those in the Chacuaco Canyon. However, other socio-cultural reasons may account for some of the differences between the bedrock ground stone samples. For the most part, watersheds seem to predict how culturally related groups will utilize resource territories. While the Western Mono and Luiseño move more or less vertically through their landscapes, along the Chacuaco, prehistoric peoples probably moved both vertically (from the floodplain to the upper canyon rim) and vertically-horizontally among the resources of individual side canyons.

Since the BGS sample most closely resembles the Luiseño model, we can interpret the seasonal movement within this side canyon accordingly. The variation in the sample indicates a variety of use but consistent push-pull behaviors in processing. It is possible that prehistoric peoples did not inhabit the canyon on a long-term basis but may have returned only a few days or a week at a time, staggering their visits based on the schedule of particular foods, animals, and ritual time frames. Results do not indicate village size settlements for the side canyon but relatively smaller, family, or extended family groups.

The lack of storage facilities and large camps may only be the effect of biased sampling of the landscape by archaeological research agendas. The comparison does not contradict the current assumptions that the people who inhabited the Chacuaco side canyon would probably have attached family or extended family territorial connections to the landscape and its resources. Indeed, understanding the camps and stone structures found in the survey area would assist our knowledge of the relationship between the BGS features and other archaeological sites.

Whether the sites were used predominately by males or females is not illuminated. Most ethnographic evidence still suggests the women of the family maintained BGS spaces. The comparison illustrates the possibility of further investigating the role of kinship and community structure through bedrock ground stone space. This type of research would help shape our understanding of past social reproduction processes and social identity formation in past contexts.

3

Landscape Context of Chaquaqua BGS Feature Sites

Bedrock ground stone features on the Chaquaqua Plateau are sometimes imagined as being ubiquitous throughout the region. The feeling derives primarily from field experiences. The case study on JE Canyon (discussed in the next chapter) further supports the impression that BGS features are found *everywhere* throughout the region, especially tucked away in the myriad side canyons. In this chapter, we examine the assumption by asking several questions. Where have BGS features been recorded? Are they all located within canyons like the JE Canyon sample? If not, in what types of landscape contexts are they found? If they are as widespread as believed, do they resemble the best-documented cases, like those in the JE Canyon survey? If not, how are they similar or different? The primary objective is to develop our knowledge of the landscape context where we find BGS features while also describing the spatial distribution of sites (along with associated archaeological and natural features) of BGS features encountered on the Chaquaqua Plateau.

Describing the spatial distribution of BGS features in southeastern Colorado is a vital first step. The spatial distribution provides a foundation for where the features are most likely to be found in terms of water, elevation above sea level, vegetation, and the availability of suitable bedrock surfaces. Spatial distribution of BGS features furnishes evidence of patterned behavior that in turn allows us to recognize the criteria that may have supported the choice of site locations. For instance, one hypothesis states that BGS feature sites are more likely to be found in the rockshelters found just below the canyon rims. Is this the case, or are sites distributed in many types of landscapes? If we rely solely on individual cases, like those in the following chapters, we might lose focus on the big picture.

There are two primary storehouses of information on Chaquaqua BGS features. The Office of Archaeology and Historic Preservation (OAHP) in Denver, Colorado, provided the GIS (Geographic Information Systems) spatial dataset. The dataset holds archaeological sites with BGS features along with information about the local environment, other features, and site context (including diagnostic artifacts). One information field stored in the spatial dataset is the "archaeological context," the field assessment of the site type. For example, a site might be designated as "a camp" or "architecture." Another information field in the spatial data set is the "landscape context." The landscape context records the field assessment of the viewshed or site's aspect. Viewshed or aspect tells us how much of the landscape is viewed from the central part of the location, often "open" or "sheltered." The dataset holds other information for exploring the context of BGS features, including the site's distance to water, elevation, and artifact analysis. Artifact analysis includes a list of tools and materials found during the site recording process.

Archaeologists at the Pinon Canyon Maneuver Site (PCMS) provided additional contextual information in excel databases for all BGS features recorded on the PCMS and Fort Carson (FCMR) in Colorado Springs, Colorado. Since base archaeologists have been instrumental in recording BGS features, the PCMS/FCMR database has information on measurements of individual BGS surfaces, not otherwise available for study. Both agencies provided access to sites or site files, which contained field notes, site sketch maps, and photographs of BGS features.

Renaud's descriptions indicate that the features occur in western Oklahoma and northern New Mexico, but do they extend north of the Arkansas River? If they do, what are the similarities and differences between the nature of BGS features beyond the Purgatoire River system on the Chaquaqua Plateau? There has been little systematic survey to describe BGS features. Their discovery is often incidental to other archaeological work objectives. Thus, although there may be a wide distribution of sites, there are many gaps between survey areas. However, synthesizing what we currently know will point out the critical gaps in our knowledge and help refine further BGS research for Chaquaqua Plateau and neighboring regions.

MATERIALS AND METHODS

Archaeological sites in the region are assigned Smithsonian numbers during archaeological field surveys. The OAHP manages the process in Denver. Completed survey forms then become part of the OAHP physical and online data repository. These data are accessible through the COMPASS system, specific to Colorado, or by request. Data sets can be anything from scanned documents to spreadsheets; or may come in formats ready for programs such as ERSI's ArcMap, a geographic information system (GIS). Research in this chapter is based only on data accessed from the GIS data set provided by the OAHP.

Other datasets were obtained from the USGS National Map. These included the National Hydrography Dataset for the Arkansas River Basin, primarily in southeastern Colorado, and included the system that drains directly into the Arkansas River and a second system that drains to the east/northeast. Maps presented here were generated using ArcMap 10.7.1.

The OAHP search perimeters for this project used historically popular terminology for BGS features such as "bedrock mortar," "bedrock metate," "fixed mortar/metate," or any other term on record which might have been used to code the presence of permanent ground stone features on a site. As mentioned in the previous chapter, terminology for BGS features is rarely consistent, so the terms used for this search are from publications, site reports, and experience. The OAHP provided excel data and GIS data formats. The PCMS excel data and their GIS data formats for their survey boundaries helped compare general patterns with other archaeological features. PCMS data supplied supporting documentation for the most current data not yet included in the COMPASS system.

To understand how BGS features fit into the broader archaeological context for the Chaquaqua Plateau, I explored four objectives: 1) feature distribution on the plateau, 2) relationship of BGS features to hydrology, elevation, and viewsheds, 3) BGS feature relationship with other types of archaeological features such as architecture, campsites, rock art, and hearths, and 4) cultural association or general chronology. The analysis only includes sites with BGS features. However, to build reliable interpretations of the BGS feature sites with other site occupations, we need to understand how they fit in with the archaeological features in closest proximity. The assumption is that the closest features and artifacts directly correlate to the BGS features themselves. For example, if a significant number of sites with BGS features also have rock art, we can explore that relationship in terms of patterns and motifs of sites with rock art but no BGS features.

A QUESTION OF EXPEDIENCY

When considering the distribution of bedrock ground stone features in the region, we expect they will be where the material exists to create them:

bedrock outcrops. However, it does not necessarily follow that they are expedient tool kits. Expedient tool kits are generally created and used quickly for a task at hand, with little time or effort invested in maintenance. Tool expedience also implies a level of convenience of availability, access, and use. Applying this concept, without explicit evidence, to BGS features only obscures their inclusion in regional archaeology.

The archaeological context of the Chaquaqua Plateau depicts a long-term (thousands of years) occupation by people who have social and technological behaviors highly resilient to environmental fluctuations and influences of neighboring peoples (Cassells 1997; Stone 1999; Zier et al. 1999). This type of tenancy underscores an intimate knowledge and relationship with the landscape. If BGS tools were just haphazard or expedient tools, we should not expect concentrations of site use in specific areas; rather, the features should be more or less randomly distributed across the landscape. Additionally, the patterns created by expedient use would exhibit little shaping (the term we use in the field to describe these features is "amorphous" or without shape). The depth would be very shallow or non-existent, as little time is invested in grinding at the surface. Further, since it is unlikely to have been used often, the grinding surface would likely have been long eroded away by natural processes.

It is possible that some of the BGS features might be *convenient* but also represent significant time investment. In modern hunter and gatherers societies, an example of this is briefly mentioned by O'Connell and colleagues (1991). The availability of bedrock space to process resources drives decisions about where to locate a campsite and how long to remain there. Regardless of the scenario, the behavioral norms associated with social life, distance to food resources and water, among other choices, are constantly being negotiated in tandem with the availability of places amenable for food-grinding tasks, thus creating a new way of looking at the regional landscape use.

BGS FEATURES ON THE CHAQUAQUA PLATEAU

There are 459 sites in the OAHP database that define some type of permanent ground stone feature. Site data for five JE Canyon sites, not yet in COMPASS, were added to this analysis. Bedrock ground stone feature sites are densest within the Arkansas River Basin. However, there are sites on the front range of the Rocky Mountains just north of the Arkansas River. The extent of BGS features spans the entire Chaquaqua Plateau (Figure 3.1). They are found along all major drainages from the front range to the New Mexico and Kansas borders. So far, the most sites have been recorded near the Purgatoire River. The density of site recording along the Purgatoire is due to ongoing archaeology on the PCMS since the 1980s.

RELATIONSHIP TO THE LANDSCAPE

The BGS features on the Chaquaqua Plateau are found primarily along the Purgatoire River and its tributary drainages and canyons. However, sites with BGS features have been recorded along many of the small rivers that eventually drain into the Arkansas River. Two central hydrologic systems drain the Chaquaqua Plateau toward the north (Purgatoire) and the east (Two Buttes and the Cimarron Rivers). The Cimarron River connects with landscapes to the west and southwest as well as to the east. The heavy concentration of sites west of the Purgatoire and north of the Arkansas rivers is probably due to PCMS and FCMR recording bias. Thus, the relative paucity of sites elsewhere on the plateau is not an accurate indication of BGS regional site density but rather the incomplete record of sites due to a lack of survey data. In all likelihood, there are many more, as yet undiscovered, BGS features sites.

The sites with BGS features fall under two primary site descriptions in the OAHP dataset (Table 3.1). The first describes the archaeological context of the site as architecture, camp, historic, or lithic concentrations. Data collected from the JE Canyon survey, not already in the OAHP database, were used as an independent category since no context exists for BGS features specifically. The BGS feature sites from JE Canyon, which belong to an interconnected ecosystem, might qualify under any or none of these sub-classifications. The second description discusses other aspects of the site that are related to landscape context, including viewshed, literally the view from the site. Open views generally have a 360-degree perspective of the landscape. Viewshed also describes the orientation of a site; for instance, an east-facing, sheltered site refers to both

Table 3.1. Summary of the Types of Descriptions Provided by Field Assessmen

Archaeological Description			*Viewshed Description*		
	N	%		*N*	%
Architecture	179	38.6	**Isolated**	20	4.3
Camp	254	54.7	**Open**	219	47.2
Historic	1	0.2	**Open/Sheltered**	3	0.6
Lithic	4	0.9	**Ranch Complex**	1	0.2
JE Canyon	5	1.1	**Sheltered**	215	46.3
Unspecified	21	4.5	**None**	6	1.1
Total	**464**	**100**	**Total**	**464**	**100**

the direction and the fact that the site has a limited perspective of the surrounding landscape. Isolated is used to define an archaeological artifact or feature with no other types of features or artifacts nearby. A site referred to as "open" has a viewshed less than a 360-degree perspective but is generally more than 180-degrees. Some open sites will have bedrock hoodoos (areas of eroded sandstone escarpments) or wooded juniper areas nearby.

Open/sheltered sites contain both 360-degree views as well as protection in the form of rock formations or heavily wooded areas. Ranch Complex is a specific type of landscape pertinent to the area during the early Euro-American occupation. Some BGS features found near ranches are associated with fences or other historic building projects. Sheltered sites are those that contain rockshelters or overhangs nearby that both limit the site viewshed while providing expedient or long-term shelter. Many types of BGS features are associated with the wide variety of sites labeled "sheltered"; these will be discussed more in later chapters.

The majority of BGS feature sites are categorized as a camp (N = 254, 54.7%), while sites characterized as having architecture (N = 179, 38.6%) are also a significant part of the sample. BGS feature sites categorized as lithic areas make up a very small percentage (0.9%). JE Canyon sites fit into any of these categories. Camp remains, stone architecture, and lithics are associated with nearly all BGS feature sites. The dense BGS sample and other archaeological features found within the JE Canyon landscape provide evidence that the JE Canyon sample is unique for the region. If considered with Metate Cave (5LA6575) and Trinchera Cave (5LA1057), the BGS elements represent the best sample for creating a regional shape taxonomy. The importance of these BGS feature sites suggests that a new category of archaeological description, one that considers the BGS feature as the central anchor, should be considered warranted.

The landscape context of regional BGS feature sites split evenly into either open (N = 219, 37.2%) or sheltered (N = 215, 46.3 %) categories. The result is surprising, considering the general impression that BGS features are primarily found within rockshelters (a sheltered context) and that they have large clusters of grinding elements, similar to those found at JE Canyon, Metate Cave, and Trinchera Cave. The pattern will be discussed more in chapter 7. However, the simplest explanation is that there appears to be a dichotomy in the types of BGS features on the Chaquaqua Plateau: those associated with open spaces and those associated with the more sheltered environments. Those in sheltered environments can be subdivided into sites with features in rockshelters or sites with BGS features in open-air but near a rockshelter. The implication for site use is important. BGS features, from those in rockshelters have a medium to a large number of elements per feature. These grinding surfaces have a defined oval-shaped basin with some depth. Those features from sites with BGS features in open-air but near a rockshelter generally have one or two grinding elements per feature, which are amorphous and have little or no depth.

In terms of the regional settlement pattern, we may have two different land-use systems. The sheltered BGS features with multiple grinding elements represent a highly intensive but localized use. In contrast, the open-air features with one or two amorphous grinding elements represent a less intensive use but make extensive use of the landscape. The ability to use exposed bedrock for a wide range of resource procurement may be one factor that afforded resilience to the people who inhabited the region for such a long time. Chapter 7 explores the distribution of these different landscape aspects to see if they represent a long-term pattern or an evolution of landscape use through time.

In the absence of direct ethnography, we can look at three landscape settings, elevation, distance to water, and site area, to better understand how past people utilized the BGS features on the Chaquaqua Plateau. Elevation provides context for the sites in terms of seasonal resource availability. Although winters can be brutal at any elevation on the High Plains, the canyons of the Chaquaqua Plateau provide respite from these conditions for all lifeforms. When the plains are covered with blizzard conditions, the canyons provide shelter from the winds, driving snow, and frigid temperatures. However, in the summers, the same sheltered areas can be subject to raging fires, flash flooding, and much hotter temperatures. If most of the BGS feature sites are associated with seasonal food processing, those in open-air contexts would probably indicate early summer to early fall use simply due to the availability of plant resources. There is little or no difference between the range of elevations for BGS features across the landscape (Figure 3.2). The elevation of open sites ranges from 4210 feet to 7000 feet above sea level, with an average of 5230 feet. Sheltered sites are found between 4177 feet to 7000 feet above sea level, with an average of 5023 feet.

The distance of a site to the nearest water source varies between landscape contexts. Permanent springs at higher elevations provide year-long water regardless of the season or climatic variation. JE Canyon illustrates the situation (Figure 3.3). A permanent spring lies within an easy walk to any of the sites. The variable availability of local water sources means that each canyon and side canyon have a wide range of plant availability and density. Thus, understanding how sites are distributed in relation to water sources is vital, not only for human use but also for other wildlife upon which humans depend. Rockshelters with BGS features (N = 27) are closer to water sources, while open sites (N = 198) and open-sheltered sites (N = 185) are farther from permanent water sources. Most BGS features that have been recorded to date are in open-air contexts. Those found in rockshelters near year-round springs nestled in more sheltered canyons make up a small number of the BGS feature sites (Figure 3.3).

The JE Canyon survey data illustrate this scenario. Localities with BGS features are found along the upper slope of the canyons, very near the permanent spring located at the head of the canyon (Figure 3.4). Sites located further from the canyon rim, in more open landscapes, are lithic tool scatters, stone structures or isolated BGS features with one or two grinding surfaces. The main difference between the open-air surfaces in JE Canyon and those in the regional sample is that the former are oval-shaped basins with depth (Lynch 2017a, b), while the latter appears to be primarily amorphous shaped grinding surfaces with little to no depth.

Lastly, the actual size of the BGS feature sites in relation to the landscape context can tell us about how people were using the landscape. Open-air sites are generally larger than sheltered site contexts (Figure 3.5). Perhaps the landscape is a constraint since there is little difference between the size of sites with BGS features in sheltered, open-air contexts or those BGF features found within the rockshelters. It may also be that the area of a site does not reflect the intensity of use but rather extensive and expedient use. In other words, open-air grinding is part of a more extensive suite of human activities, such as foraging, hunting, or tool acquisition and production.

ARCHAEOLOGICAL CONTEXTS

The archaeological context situates the bedrock ground stone feature among the other features or artifacts included in the site description. When describing the context of a site, archaeologists have several descriptive choices, for instance, rockshelter, stone architecture, hearths, or lithic scatters. Some site contexts contain a combination of descriptors. These coding procedures support the development of a theory of how a site was used through time. It allows us to search for patterns of archaeological features and contexts associated directly with BGS features. Using data from the OAHP, three main types of features are associated with BGS features (Table 3.2): rockshelters (32%), stone structures or circles (28%), and rock art (23%). There are three minor categories of features: lithic scatters (meaning the presence of stone tools or evidence of making stone tools, 11%), tool grooves (2%), and lithic quarries (.4%). These categories may overlap. For instance, a BGS feature site might be described as a rockshelter site with stone structures, stone tools, tool grooves, and rock art. These data refer only to features found within specific site boundaries.

Table 3.2. Summary of Archaeological Features Found within BGS Feature Site Boundaries

	Rock Art		*Rockshelter*		*Tool Grooves*		*Stone Structures or Circles*		*Lithic Scatters*		*Lithic Quarry*	
	N	%	*N*	%	*N*	%	*N*	%	*N*	%	*N*	%
Present	107	23	149	32	10	2	132	28	49	11	2	0.4
Absent	352	76	310	67	449	97	327	70	409	88	457	98.5
Unknown	5	1	5	1	5	1	5	1	6	1	5	1.1

Rockshelters and stone structures are important from two perspectives: they imply shelter or protection from the elements for human inhabitants and suggest the possibility of storage space for food or other resources. A rockshelter at 149 sites of the 464 sites in the sample does not mean that BGS features are found within the rockshelter, only that a rockshelter is nearby. The bedrock grinding surfaces are near the rockshelter in the open air. So, the absence of rockshelters at 310 sites with BGS features and the absence of stone structures at 327 BGS feature sites strongly suggest that open-air grinding spaces dominate the archaeological context. How are these open-air sites being used? Are there camps nearby that are not yet recognized? What are the similarities and differences between open-air and sheltered BGS features?

Ethnographic examples of how grinding is done in open air might be helpful. Gayton (1948a:64) depicts temporary shelters built over bedrock mortar features to shelter the person while they work. These types of temporary shelters would leave little trace evidence in the archaeological record. A second image from modern hunter-gatherers (O'Connell et al. 1991) shows the grinder working without shelter in open-air along with her child. The context of the grinding scenario is that of a functional space near a campsite without permanent structures. In Gardi and Gardi (1995), young grinders also work without shelter. However, the archaeological context is within the environment of a small village. Therefore, the archaeological context of BGS features suggests two types of use, near rockshelters or permanent structures (or both) and those found in open-air contexts. Is there a difference in the seasonal use or symbolic nature of BGS features?

Rock art on sites with BGS features or the BGS features themselves may provide evidence of the latter. In the JE Canyon survey, BGS feature sites with a high quantity of BGS elements are often associated with pecked rock art motifs, a pattern repeated at Metate Cave and Trinchera Cave. The JE Canyon sample has both intensively and lightly used BGS features. Intensively used BGS features are those with more than ten grinding elements with defined shapes and depth greater than zero. These sites are *always* associated with large complex rock art panels and small-scale panels with abstract motifs. In general, most of the rock art found in the JE Canyon sample was in direct or very close proximity to BGS features (Loendorf 2008; Lynch 2017a, b). Metate Cave and Trinchera Cave also have complex rock art panels in association with intensively used BGS features. The relationship between grinding and symbolic use of space as represented by the rock art panels is an exciting line of inquiry that has yet to be explored. The two types of features may have been used together or may represent a multi-layered use of space in time or in the seasonal lives of the people who created and used the sites.

Other nearby types of features such as lithic scatter, quarries, and tool grooves occur infrequently. Lithic manufacture or quarrying is only associated with 11 percent and 0.4 percent of the BGS feature sites, respectively. The small portion of lithic manufacture sites nearby potentially expedient grinding sites (those in open-air with amorphous shapes and little to no depth) could be a sampling bias. An example of this comes from the JE Canyon survey, where the grinding activities were far enough (more than 50 meters) from activities of lithic manufacture that they are categorized as a separate site. Ethnographic data indicate it is unlikely that food processing sites would be completely isolated or away from other socio-cultural activities on the landscape (David 1998; Gardi and Gardi 1995; O'Connell et al. 1991). We should expect to find campsites or habitation sites nearby. In the next chapter, JE Canyon Ranch data suggests that site-level analysis biases data interpretation, thus obscuring different sites' connectivity.

Tool grooves are associated with making tools from bone or other materials (Feyhl 1980) but may also represent symbolic behavior. Tool grooves generally exist as panels of related grooves. However, they may occur as isolated elements (see Lynch 2017 a, b, site 5LA11451). Tool grooves are present at only a small number of the sites in the regional sample (N = 10, 2%). Metate Cave and 5LA11451 (in the JE Canyon sample) have large panels of tool grooves directly associated with the ground stone workspaces. In the JE Canyon sample of eleven sites with BGS features, only one has tool grooves. In conjunction with BGS features and rock art panels, the rare presence of tool grooves may represent a place with "special" meaning to the inhabitants.

Artifact and Material Associations

Although there are few lithic features associated with BGS feature sites, the presence of debitage (N = 232) at 50 percent of the sites indicates tool reduction was occurring at these sites but at a small scale. The presence of chipped stone could indicate on-site tool reduction or may represent site erosional processes. The latter example was noted at 5LA11455 in the JE Canyon survey, where a medium-scale lithic scatter (more than 100 chipped stone or broken tools) was located on the terrace directly above the BGS features. A few flakes had washed into the site from this upper terrace, so while the grinding and lithic reduction area were not in the precise vicinity, they were very close neighbors. The two sites were separated by the natural landscape terrace of about 30 meters, so they should have been recorded as two different sites. However, the clear evidence of erosion from a higher terrace to a lower terrace allowed us to consider the two as a single, vertically oriented site with two distinct work areas: one for grinding and one for lithic manufacture.

Not surprisingly, the presence of manos and metates occurs at 34 percent of the BGS feature sites (N = 159), although this number is smaller than expected (especially of manos or pestles). These portable tool kits are often cached or curated for use and stored elsewhere. The presence of such tools may represent a type of local storage for future use instead of being related directly to the BGS features. They may also represent two types of work processes or shifting practices of grinding. Since portable tools are curated, they may be out of context.

Lithic tools are found at many BGS feature sites (Figure 3.6). Projectile points (N = 120), bifaces (N = 112), cores (N = 108), utilized flakes (N = 98), and scrappers (N = 72) make up the dominant tools' categories. Only four tool categories differ between the primary archaeological contexts (Figure 3.7). Debitage, or the remains of tool manufacture or shaping, is mainly found in sites categorized as camps (as opposed to architecture). Those sites found in or near rockshelters have a higher amount of tool-making refuse than sites categorized as rock art sites or those with stone structures. While the remnants of other types of ground stone tools (manos and metates) are found at all site categories, these are often found within camp and rockshelter contexts. Other tools such as projectile points, bifaces, utilized flakes, and scrappers are found across all archaeological contexts.

Stone Material at BGS Feature Sites

The lithic material assemblages are dominated by local quartzite, chert, and basalt (Figure 3.8). Jasper and alibates are much rarer (Figures 3.8, 3.9). While the sources for these materials are not discussed in detail, the closest quarries, at the head of the Arkansas, Purgatoire, and Huerfano rivers, maybe the closest sources for quartzite, basalt, and argillite (Black, 2000). Jasper, chalcedony, and petrified wood are available on the Front Range (Black, 2000). Alibates may originate in the Texas Panhandle.

Distribution of the more exotic (relative to local quartzites and basalts) materials found at BGS feature sites occurs across the archaeological contexts (Figure 3.9). Chalcedony (N = 40), silicified wood (N = 35), and obsidian (N = 32) materials are relatively common at rockshelter and stone structure sites. Jasper (N = 6) and alibates chert (N = 8) are found at fewer BGS feature sites. Although BGS features are found at about the same frequency in open and sheltered contexts, the non-local materials occur more frequently in open-air contexts (Figure 3.10). Chalcedony is found at more sheltered sites than silicified wood and obsidian. These materials rarely occur together although, chalcedony, Alibates, and obsidian are found together at 5LA4751, an open-air site with five BGS surfaces and tool grooves. Alibates and obsidian occur together at 5LA8757 and 5LA12616. The first is part of the Trinchera Cave historical district, while the second is described as primarily a "cache site" with a single, highly weathered BGS feature. The combination of jasper, chalcedony, and silicified wood occurs only at 5EP6069, along Turkey Creek, Colorado. The site is part of a cluster of sites that may be part of a game drive complex. While the site is rich in lithic artifacts and materials and has portable ground stone tools, the BGS feature is highly eroded.

Two BGS feature sites are lithic quarries. The first, 5LA2280, has a long-term occupation from about 5000 BP to 225 BP, mainly based on the diagnostic materials. The quarry is associated with quartzite, although the site has many material types, including obsidian. There are several thermal features in addition to the 34 BGS features, all of which are typical of open-air context grinding features. The second, 5BA356 (located close to the Colorado-New Mexico border), has multiple BGS features. The site is mainly discussed as a sheltered camp where tool reduction occurred.

DISCUSSION AND CONCLUSION

Bedrock ground stone features sites found throughout southeastern Colorado occur in two landscape contexts, open-air and sheltered. The main differences between the contexts are the shape and depth of the grinding elements. The archaeological contexts of BGS features are within or near camps (these can be open-air or near sheltered areas, including rock-shelters). The grinding features are also frequently associated with stone structures. These structures vary from stone circles to adjacent wall structures associated with the Apishapa culture. While many features appear to be associated with short-term occupations, some sites are associated with long-term occupation or re-occupation.

The seemingly pervasive pattern of two types of BGS features sites, open-air and sheltered, has to be further explored.[1] In the next section, the methods used to describe and interpret BGS features on the Chaquaqua Plateau are delved into using three site locations found in rockshelter contexts (Figure 3.11). The first is JE Canyon, a cluster of sites in a single side canyon with open-air and sheltered sites. Sites with BGS features in the Chaquaqua Plateau were chosen for the presence of appropriate materials for grinding activities and their proximity to water and resource rather than their proximity to shelter. The second case study in methods is the Metate Cave site on the Pinon Canyon Maneuver Site near Trinidad, Colorado. The last case is that of Trinchera Cave, on State managed lands near Trinidad, Colorado. Each case illustrates different recording methodologies as they have evolved in the region.

NOTE

1. Exploring the reason for this signature is part of a new investigation. The first phase is to analyze the actual field signature against other types of use wear signatures (for instance, portable ground stone from the Southwest or from other Plains sites). A second experiment is underway that explores the morphometric signatures of natural erosion and open-air basins from JE Canyon.

II

METHODOLOGY

4

Capturing the Fundamentals

JE Canyon Ranch

Recording methods for bedrock ground stone features have typically followed protocols developed for portable and fixed metates in the southwestern United States (Adams 2002; Bartlett 1933; Eddy 1964; Woodbury 1954) or bedrock mortars in the western United States (Dill 1975; McCarthy et al. 1985; True 1993; True et al. 1974; True et al. 1991). The logic behind using one or other of these methods relates to our assumptions about the function of BGS features within the daily lives of ancient people, namely that the features were used to grind food resources. The California model seems most applicable because of the similarities in the archaeological remains (see chapter 2). The BGS features are not portable materials but represent a human-constructed space within a specific ecological area. Thus, we assume a relationship between what was being worked or processed (local food resources) and the places created on the landscape itself (ground stone workspaces where food is symbolically or functionally associated with cultural behaviors). Additionally, some of us wonder if the features are long-term, habitually used areas such as those in the western United States (California, particularly) or whether they were low-intensity use areas that may have served only temporary and short-lived purposes.

Spatial analysis of the BGS features on the Chaquaqua Plateau (chapter 3) suggests two main archaeology patterns. Features found within rockshelters or near them and those found outside of rockshelters, in the open air. Unlike the JE Canyon sample used for comparison in chapter 2, the variations between the landscape contexts seem to illustrate that some

BGS features may have been used as expedient tools. In contrast, others were revisited and utilized repeatedly through time. Some open-air contexts may have been habitually occupied, but the lack of depth on the ground stone surface coupled with ill-defined shape suggests short-term sporadic utilitarian use.

This chapter examines the traditional pen and paper methods used to describe differences in morphology for BGS features in both types of landscape contexts but found close together in a bounded landscape.[1] Of particular interest are any patterns in the material objects that support interpretations about how the feature may have been used or how the site may have been occupied. A bounded landscape facilitated the comparison of feature types, their ecological setting, and their relationship to one another. As noted in chapter 3, many of the open-air features are typically located at a distance from other types of ground stone sites. The best way to analyze or interpret amorphous, shallow ground stone elements is still being developed.

ENVIRONMENT OF THE BOUNDED LANDSCAPE

The survey area was part of an unnamed side canyon of Chacuaco Creek Canyon—a large watershed that flows north until it spills into the Purgatoire River. The canyon is part of the Raton Section of the High Plains, within a physiographic sub-region designated by Fenneman (1931:37) as the Chaquaqua Plateau and described as "a stratum plain trenched by canyons." Although Fenneman (1931:43) used an antedated spelling "Chaquaqua" to designate the stream, the canyon, and the incised plateau, USGS 7.5 topographic quadrangles record the names of the stream and canyon as "Chacuaco."[2] To avoid confusion, the name Chaquaqua refers to Fenneman's physiographic sub-section of the High Plains because it is still widely used and appears on many regional maps. Chacuaco refers to the drainage system that includes the stream, watershed, and canyon, appearing on all modern USGS maps.

The Arkansas River, which appears to mark the northern extent for the intensive use of BGS features, marks the northern boundary of the Chaquaqua Plateau. The southern portion of the Chaquaqua Plateau includes lava-capped mesa formations that dominate the landscape but terminate at the Mesa de Maya formation. The physiographic change to the grasslands just east of the Colorado border forms the eastern boundary. The western boundary lies at the foothills of the Sangre de Cristo Mountains. Some authors, including Fenneman (1931:37), subsume the Chaquaqua Plateau into the Las Vegas or Park Plateaus. For these investigations, the distinctive drainage systems of the Chaquaqua Plateau afford abundant environments for game and food gathering, thus represent a more suitable physiographic section to describe related BGS features.

Chacuaco Creek is an intermittent stream originating west of Mesa de Maya, about 40 km south of the project area, near the Colorado border. The stream drains an area of about 424 m^2 (von Guerard et al., 1987:33) before finally merging with the broadest part of the Purgatoire River to the north. The dynamic canyon system is dotted throughout, with naturally occurring springs (Fenneman 1931:43–44). A riparian flood plain characterizes the lower elevations of the modern Purgatoire and Chacuaco drainage systems. The terrace slopes, rising between 60 and 270 meters above the floodplain, are covered with grasses, cacti, shrubs, and juniper-piñon ecozones. The upper canyon rims are typified by a mixture of pinon-juniper and open plains vegetation, but at the head of many, unnamed, tributary side canyons, especially near the permanent springs, there are a variety of trees, shrubs, wild fruits, grasses, and nuts (Shaw et al. 1989).

The unnamed side canyon surveyed drains into Chacuaco Creek Canyon after merging with Bachicha, Tobe, and Poitrey Canyons. At this juncture of canyons, the Chacuaco floodplain has riparian vegetation and extremely steep canyon walls that rise more than 150 meters above the floodplain. Vegetation on the upper rim is a mixture of open plains with juniper-piñon woodland ecozones. Plants observed at the intermediate elevations (around 5299 feet above sea level) include oak (*Quercus gambelii*), pinon (*Pinus* sp.), and aspen (*Populus* sp.). Wild plums, various grasses, and cacti cluster near the permanent springs, regularly occurring stands of water, or the permanent seeps within local rockshelters. At least two rockshelters (5LA11451 and 5LA11455) have heavy seeps that support ferns within the sandstone cracks. Shallow rock overhangs and rockshelters have formed on the Dakota Sandstone escarpments in the recesses of the small side canyon. There are eleven sites with BGS features in this area; more than half of them overlook the Chacuaco floodplain, the others are in more secluded recesses of the canyon.

RECORDING BGS FEATURES: BEST PRACTICES

Recording BGS features is not a standard practice in southeastern Colorado, which can be problematic when archaeologists attempt to access metric data from standard databases like those from the Pinon Canyon Maneuver Site (PCMS) and Office of Archaeology and Historic Preservation (OAHP). Along with the challenge of dating BGS features, they are not well-studied due to the lack of consistent data sets and the irreproducibility of some of the data already collected. To provide an example of the best practices in recording, I describe a typical field recording process that will result in solid data sets that can be used to reproduce (or question) results and build a more comprehensive discussion of these types of archaeological features.

Earlier chapters have covered the minimum types of data necessary to expand our discussion of features and sites. These are the dimensions of the planar surface of individual grinding basins, the actual placement of the grinding basins on the feature, and the feature's location within the site. Once these minimum elements are described, comparisons can be expanded to include nearby sites. The first step is to create a detailed map of the site, followed by scaled sketch maps of every feature on the site. Each of these drawings has to explain the feature numbers and the individual identification symbols assigned to each grinding basin. In the following pages, the latter are referred to as grinding elements or just "elements." Measurements of individual elements should be referenced on the feature drawing.

In terms of the individual elements on BGS features, it is not enough to just record the length, width, depth, orientation, and slope of each basin. They must be identifiable as specific elements on the site. The author, trained by Lawrence Loendorf in rock art recording, adapted his field practice. In this method, drawings supply a planview illustration, while supplemental recording forms provide a complete description of all elements on the feature. Photo documentation of the features, while imperative, is not enough to replicate data nor to verify use-wear patterns or field assessments.

The methods described below are time-consuming processes that may or may not fall within the time frame of the archaeological project design. For example, in the instance of fast-moving reconnaissance projects, there will likely be no time for the data sets to be recorded, so photographs have to ensure that the feature(s), along with all the grinding elements, is (are) viewable. A scale and a north arrow should also be in the photograph. The goal is to provide enough information for researchers to decide whether the site warrants additional study. Most, but not all, of the reconnaissance forms on file with the OAHP provide excellent written and photo evidence. Although not incorporated in regional morphometric research, reconnaissance survey data add to our understanding of the spatial distribution of BGS features sites in the region. The data assist in prioritizing which BGS feature sites need additional work to improve the regional archaeological knowledge.

In other types of cultural resource management (CRM) surveys, or archaeological research projects, the data gathering process for BGS features should be planned for and included in the research design. The basic, manual recording methods described below do not require additional skills-sets or equipment. The method only requires a dedicated person or persons to complete the work in the field as well as the follow-up work in the lab. Those data can help us better understand the relationship of the BGS features to the other archaeology sites, in addition to aiding archaeologists to interpret the nature of human behavior on a site or within a cluster of sites.

A brief note about photography is warranted. In the next chapter, I will discuss collecting data sets to develop 3D models of BGS features. While these will produce models that multiple researchers can use to replicate any set of measurements, they do not obviate the need for detailed field documentation. The pros and cons of 3D modeling methods will be discussed in the next chapter. Suffice it to say that a field planview map should accompany photographic methods no matter how basic. In the lab, field notes supply details to support the model-building process. While it is possible to assign or reassign elemental identification numbers in the lab, a good field drawing provides a complete record of what the archaeologist observed at the moment. The notes are essential, specifically when multiple people are recording the features. Providing robust information about BGS features is a concern that goes to the heart of scientific visualization, so it will be discussed in detail here and in the subsequent chapters.

FIRST STOP: THE SITE MAP AND PLANVIEW FEATURE DRAWING

Every project has challenges staying up to date on the latest in technology. However, many archaeological projects may only have time and money for rapid pedestrian surveys designed to provide quick information about cultural resources in a particular area. The OAHP COMPASS database has many records from these impressive surveys. All incorporate at minimum a photograph and brief verbal description. It is unlikely that these surveys, which play a vital role in understanding the extent of cultural resources, will have time or energy to bring digital recording devices (360-degree cameras, total stations). However, work on JE Canyon was very low-tech. All field recordings used a hand-held GPS (to supply a datum but also to map site boundaries), a compass, graph paper, pencils, and various sizes of tape measures. These planviews, when compared with 3D models, proved very accurate (Lynch et al. 2017). They had the one commodity that many rapid archaeological surveys lack: time. The methods described below can be worked through quickly, with practice, in the field.

The JE Canyon survey took place in two phases (2008 and 2011). The first was a rapid reconnaissance of the property by the Colorado Rock Art

Association, who had consulted with the property owner on the best sites to invest recording time to move quickly through the canyons surveying for rock art. The number of rockshelters with BGS features presented a problem for the team. Lawrence Loendorf created a ground stone recording team (myself, a scholar from Argentina, and two CRAA volunteers). One team was dropped at BGS feature sites while the rest of the team proceeded to survey. The independent team had one day to develop scaled site maps and feature maps. Dr. Loendorf kept in contact with us through radio but would come by during breaks if the survey team happened to be in the neighborhood. These quick field drawings (Loendorf 2008) provided the basis for work completed in 2011 when there was more time to revisit sites for comprehensive data collection. The rest of this chapter discusses the work completed at one site, 5LA11446 (comprehension survey data is in Lynch 2017a, b).

THE SITE MAP

We drew two planview sketches at 5LA11446: an overall site map and a rockshelter map. We drew the site map from a base grid of two 70-meter measuring tapes laid out on a north-south/east-west orientation at the center of the site (box lines on Figure 4.1). The objective was to capture the relationship between the impressive hunting motif rock art panel, the fire-cracked rock features, and the rockshelter with the bedrock ground stone features. The author had the most experience with drawing rapid, scaled site maps. Hence, the second recorder, a rock art specialist, took over the documentation of the ground stone basins within the rockshelter.

We established a working datum at a central location. The site was on private property but now is part of a conservancy group. Our agreement with the property owner, at the time, was to be as non-destructive as possible, so the datum was an arbitrary point at the center of the site. The flexibility allowed us to set up two grid lines for the maps, one to determine the overall site area, the other to capture details of the rockshelter. The significant rock art panel had already been recorded (Loendorf 2008). It is a dominant aspect of the site, so there was no need to coordinate our temporary system to the previous recording. Once the grid lines were set up, we used a scale (1 cm = 1 meter) to represent as much of the site as possible and used a hand-held compass to square the grids. It is a simple field method that produces very accurate results.

We determined the site boundary as the limit of main features, although the landscape provided some natural boundaries. For instance, the terrace above the site is separated by a sheer boulder face (Figure 4.2). The site's topography tapered quickly into a steep, colluvial slope (more than 35 degrees) of sand, pebbles, and sparse grass clumps that dropped at increasingly precipitous intervals into the floodplain about 300 meters below the site. So, while there are other features near this site, they were beyond 25 meters of the site boundary. We recorded them separately, mainly using our hand-held GPS units. We used a hand-held measuring tape to plot the site features on the map from the central grid axis (Figure 4.1). We used an arbitrary 10 cm to take measurements of artifacts or other features on the site. Of course, the process varied for relevant details. For instance, juniper trees, boulders, and the rockshelter were measured at multiple points from the central grid lines regardless of the 10 cm standard.

We drew in the rockshelter outline then moved the second grid line to add the details of the rockshelter to the site map. We interpreted the relationship between the BGS features within the rockshelter and the rock art panel as the most important detail to capture. However, we drew in the other key features within the shelter (a stone wall), the shelter dripline, and the rockshelter profile. Grinding processes are often associated with ritual grinding events such as first fruits ceremonies or social rites of passage (for example, puberty rites and marriage rites, for more on these events (see chapter 8). If grinding features are associated with such events, the location of BGS features to one another within a site context is critical to understanding how ritual processes or ordinary grinding relates to the workers themselves (for example, how many grinders could fit in the space or whether BGS features are similar or different thus perhaps representing different periods of grinding or functions).

The shelter profile and dripline are vital pieces of information that increase our ability to measure social aspects of grinding but also help us understand the potential erosional aspect of the grinding elements. The shelter profile was mapped along the gridline (refer to the profile inset of Figure 4.1 marked A′ to A) using a line-level attached with blue tape to the back of the shelter. The line-level followed the east-west grid line. We took measurements from the profile line at 10 cm intervals to the ground and the ceiling (see profile inset on Figure 4.1). The dripline procedure used a string with a plumb-bob or weight attached to the end. Standing directly under the ceiling of the rockshelter where it begins to slope upward, the recorder used the weighted string to obtain accurate points in space then drew these on the map. Mapping a rockshelter with two people can be done quickly; one person measures the point on the floor of the rockshelter while the other person holds the weighted line. However, one-person mapping produces accurate measurements, although it is a bit more time-consuming. Dripline measurements were taken at about 30 cm. The interval depends upon the characteristics of the rockshelter.

The mapping goal of the CRAA 2008 survey was to record, with as much accuracy as possible, within one day (6:30 am to 6:00 pm). Both recorders had extensive experience drawing expeditious, accurate archaeological maps both for sites and for rock art. These types of sketch maps could be completed in much less time, depending on the project goals, nature of the site features, and the experience of the field recorders.

BGS FEATURE SKETCH MAPS

Developing scale drawings of the features is as important as taking good photographs of the features. Photographs are good evidence, but as illustrated in chapter 7, the amorphous features located in open-air contexts are tough to capture. If another researcher cannot see the proposed usewear, there is no way to verify the field observations without revisiting the site. Although the photographs should aid us in relocating the features, this is not always the case. The nature of the open-air features and their relationship to the other BGS features or the landscape must be further investigated to better understand how grinding features fit into the existing archaeology. Further, we must develop robust methods to distinguish human agency from natural processes. Good field sketches, which admittedly take some time to draw, are necessary to understand whether morphometric analysis can help understand the nature of the open-air features.

In the JE Canyon survey, the crew documented all features categorized as BGS. In the 2008 survey, it became apparent we would not have time to record all of the sites to the best level of detail, so field recorders drew rapid sketch maps. These maps seldom identified all the grinding basins on the features, but they illustrate how field observers can quickly assess important aspects of archaeological features. The rough sketch maps also provide a baseline for field observations. The process is the same as described above, designate a team or teams to learn the basic recording skills for sites and BGS features, then turn them loose with a defined time frame. Even if all basins are not captured, other researchers will be able to identify the features from the main grinding elements. More important, they will be able to know which grinding elements belong to which measurements. The importance of the procedure will shortly become more evident.

The process of recording BGS features should be done in three phases: 1) feature sketch map, 2) feature recording form, and 3) good photographs that capture the feature within the environment. The photographs should show the position of the feature within the broadest landscape possible. The first should show the immediate environs. The second provides the feature overview that clearly shows the grinding element(s). North arrows and a scale provide additional context for what the archaeologist is observing. The time frame for survey work might have to be extended to account for the process. In reconnaissance surveys, there will not be enough time for all of these steps. A quick sketch map with notes about the measurements of the main surfaces can be made quickly, and the recorder can snap two photographs with a scale bar and north arrow, noting the position of the camera shot on the map. In rapid surveys, there may not be time to develop a site map, so just adding general boulder measurements and ceiling height as well as an idea of the dripline (if the feature is in a rockshelter) will help future researchers. If the feature is in the open air, any vegetation should be noted. The recorder should describe the general slope of the features on the ground and their field observations of the types of natural erosional processes in the immediate environment.

When time exists to record the surface in detail, the quickest low-tech method starts with the north-south/east-west grid line, as with the site map. In the 2008 survey, the feature grid line was set up at the beginning of the recording process. In the case of 5LA11446, three distinct grinding features were within the rockshelter. Blue tape was added to each element with the feature number and individual field identification number. The person with the most experience with BGS features should make the identification. In the case of little experience, good photographs and notes can clarify any issues.

Once the element and feature identification are complete, the ground stone recorder can start filling out the forms and draw the feature sketch map. At 5LA11446, the site mapping went relatively quickly because we worked as a team. One recorder sighted spatial markers, making sure the tapes were accurate and calling out the metrics. However, the overall site recording process was the same. The site map was drawn first, then the rockshelter map, finally, the individual elements were recorded. The process redundancy is important because it gives the recorders a chance to become familiar with the big picture. As with recording rock art, it is essential to check for the best visibility to record the features, especially amorphous grinding elements.

Feature sketch maps can be easy to draw. At 5LA11446, one person quickly drew the feature map, while the other completed BGS feature forms and measured the elements. In the 2008 survey, the recorders split up in the rockshelter, mainly because of space. One recorder filled out the paperwork; the other drew feature maps. Bedrock ground stone feature maps were drawn one at a time, using a tape measure to plot each grinding element from the rockshelter grid line. Four coordinates were plotted: two on the longest axis and two on the minor axis. The scale for each feature map was chosen to get the most detail onto the sketch map. Again, time was not an issue. However, sometimes the scale will be a problem. Improvised methods are acceptable as long as the longest axis and the widest axis are correctly obtained, and the improvised method is appropriately noted on the feature forms and sketch map. These drawings provide an understanding of the spatial distribution of the grinding basins on any given feature. At a minimum, they provide the means to connect the measurements (collected on the feature forms) with the real world through the site map. Details are important, but precision can be adjusted if the need exists, and the decisions made clear in reporting and analysis.

Reproducibility and transparency of research are so necessary that it is worth repeating this point: the primary importance of the BGS feature map is to connect field measurements with real-world objects. The map, as scientific visualization, provides the best manner of creating replicable data for analysis of any or multiple research questions. But what are we measuring? Why do we need to do this in a reconnaissance survey?

The measurements (major axis, minor axis, and depth) of BGS features and their grinding elements are the first step to understanding any number of questions, from simple descriptions to broader site and regional comparisons (as seen in chapter 2). The longest axis (also called major axis) or element length and the widest axis (also called minor axis) or element width, together, illustrate the grinding motion, the length of the movement (perhaps even the extent of the grinder's reach), and potentially the size of the tools used to make the pattern we are observing. As illustrated in chapter 2, simple calculations of shape and size, using these two basic linear measurements, allow the mathematical comparison of shapes that may appear visually to be the same. We can compare variation within-site or between sites within a landscape (Lynch 2017a, b). The third critical measurement is depth. This measurement has proved to be the most error-ridden variable between observers but still offers insight into the basic shape, particularly how long each basin might have been used (Lynch et al. 2017). The latter has implications for how long sites may have been occupied.

RECORDING FORMS

The recording forms for BGS features were developed from the Colorado OAHP rock art recording form and the Pinon Canyon Maneuver Site feature forms. The goal of the form is to capture metric data for every grinding element on every feature and capture data relevant to the location, the time of recording, and general recorder impressions. Information gathered in the first part of the form is related to the feature's location within the site. The feature number should be in sequence with the site inventory of features and should be explicitly noted on all planviews.

A basic understanding of the parent rock type provides information on feature orientation and design potential, which can be used to understand user needs. The illustration of the physical layout helps us explore how people may have dwelled within the space. While not necessarily quantifiable, recording observer impressions provide qualitative information that helps build an impression of the site and the recording conditions. The qualitative information is particularly useful when no longer in the field. Reflective descriptions of the recorder's experiences while filling out paperwork helps the analyst not only find errors but also to design the quantitative analysis. For instance, the hypothesis that there may be sub-features, clusters of grinding elements that represented group work or multi-step processes, was proposed by Mary Lou Larson (personal communication, 2011) after discussing the relevance of recording field impressions.

The recording form also includes an area to record all the grinding elements and their measurements. Blue tape or other non-invasive markers placed on each ground stone basin links the feature forms with the feature sketch maps as well as the site map. The form also records ceiling height measurements over each ground stone basin to better understand the internal design of the features.

One of the more subjective parts of the form is the qualification of the observed shape by the observer. Is there a shape? If so, is it an oval or a circle? The regional ground stone archaeology is still in the descriptive period. We are still describing the features and the grinding basins to refine regional morphological patterns and their interpretations. The field recorder's impressions of shape add a rich description (it is genuinely an impression) that should exhibit a high amount of variation depending on the number of grinding basins encountered and field recorders. However, most sites end up with a subset of only a few descriptive terms: teardrop, egg-shaped, oval, circular, rectangular, amorphous (generally applied to eroded basins), triangular, and elongated oval.

As we saw in chapter 2, these subjective descriptors were less predictive of actual morphology than mathematical shape descriptors, especially when comparing shapes between cultural regions such as the western United States and the High Plains. A forthcoming study applies the same regional analytics to the open-air sites recorded on the Chaquaqua Plateau to the JE Canyon sample to determine how reliable these mathematical shape estimates are for categorizing similarities and differences between open-air and sheltered sites.

Completing the field recording form is the most time-consuming part of the recording process, so it may not be feasible for rapid reconnaissance surveys. In the 2008 JE Canyon Ranch survey, we recorded quite a bit of detail in one day (about 11 hours), primarily because of necessity. A good data set was needed to write grant and dissertation proposals. The 2008 expedition was the only chance to collect enough data to assess the viability of additional fieldwork.

In 2011, fieldwork on the BGS features in JE Canyon resumed. The research design called for regular nine-day field sessions between March and May, which offered more time to collect data and test out multiple data collection methods. It also allowed time for intensively surveying the areas between known BGS features sites within the same side canyon. Dr. Loendorf, who was more familiar with the canyon, took the lead for the additional survey, while the author took the lead on the BGS features recording. Together, we collected detailed drawings and feature data for all previously located sites and the newly discovered sites.

We were in the field for March, April, and May, working twelve hours days, with about three hours of lab work in the evenings. Volunteers from the University of Wyoming and the Colorado Rock Art Association (CRAA) joined the April and early May fieldwork. The project also collected data for an inter-observer comparative study (Lynch et al. 2017), collected additional pollen samples (Lynch et al. 2013 and forthcoming), and collected photogrammetry data (Lynch et al. 2017). May fieldwork was dominated by fact-checking sketch maps and recording forms to ensure that they were as accurate as possible so that the grinding elements on the feature sketch maps could be relocated in the real world by independent observers.

ANALYSIS

Once back in the lab, data from the recording forms were entered into an excel spreadsheet, including descriptive comments. Field forms were typed up, and plan view maps were traced, then scanned to created digital files that could be touched up in Photoshop and Paint. The digital files are stored at the University of Wyoming while awaiting archiving with the OAHP and the Chacuaco Ground Stone Project.

Work focused on two research objectives. The first objective was to determine the best methods to document these types of features to be included in regional analyses of prehistoric material culture. The second research objective explored how these features provide insight into how the ancient people living in the canyons incorporated BGS features into their social lives and seasonal routines. The latter research objective requires understanding chronology (discussed in chapter 7) as well as an understanding of how these types of features have been incorporated within social reproduction processes or knowledge production in other grinding cultures (discussed in chapter 8).

Once the descriptive data have been collected and processed in the lab, they have to be developed into meaningful information. The first step considers the similarities and differences in the shape and sizes of the features. The reasoning is that a single person would create similar shapes and sizes for similar processes. So, variation indicates either multiple people working at the features or different grinding processes or resources. Shape analysis and morphometrics is the method best suited to understand functional aspects of grinding elements used to illustrate social life and the relationships of BGS features to other landscape use strategies.

Shape Analysis and Morphometric

Shape analysis for BGS features always begins with three central measurements taken on any grinding element: length, width, and depth. The length is the longest measurement possible between any two points. The longest aspect of the grinding surface indicates the motion or directionality of the grinding movement (regardless of what type of resource is being ground). In oval or ovoid shapes, the length tells us how far materials were pushed then pulled back along the surface area. We refer to it as a push-pull motion. The length of the grinding element is determined by points along the planar surface, not within the basin itself. These points can be obscured by polish around the grinding basin's edge, accounting for some statistical error between different observers. The grinder or worker would probably have sat at one of the ends of the longest axis if they are ovoid or rectangular shapes. Arguably the length of the element provides information about the length of a worker's arm and might indicate whether the worker was sitting or kneeling while grinding.

A standard metric measuring tape was used to record the long axis to the nearest millimeter in the field. As discussed earlier, locating the boundary of the use surface is subject to observer bias, especially for shallow BGS elements and amorphous ground surface areas. If any element rim was eroded to the point where one of the points along the long axis could not be identified, no measurement was recorded.

The width, also measured in millimeters, is the widest part of any given element. The other term used for this linear measurement is the minor axis. It bisects the major axis on the planar surface. The perimeter of the planar surface of the grinding area approximates the basin size and shape. The minor axis evidences the lateral span of reciprocal (push-pull) or rotary movement within the grinding stroke.

In most cases, the width reflects the size of the ground stone tool used to work on the surface (for example, a one-handed mano, two-handed mano, or pestle). It is also possible that the user intentionally designed the shape of the grinding basin based on non-functional criteria. Without more information, inquiries along this line remain speculative.

The depth of a grinding element refers to the deepest part of the basin. Taking manual dimensions of this linear measurement is fraught with imprecision. Without special equipment, the best tool to quantify depth is a mechanical caliper. The recorder must locate the "deepest" part of the grinding element, place a ruler or other thin, sturdy material across the minor axis, extend the caliper to the bottom of the basin resting the base along the ruler. The process can be done with a metric tape measure as well. As long as the same person collects the measurements using the same method, the measurements are helpful for generally comparing depth. Volumetric analysis should not be attempted using this method.

Morphometric analysis, or the analysis of the metric shape of a grinding element, includes developing a means to mathematically describe the element's shape using the length and width and determining the relative size of the elements. Length and width offer a rough estimate of the grinding surface; however, they do not provide an understanding of the shape of the grinding element. New variables were developed using length and width. The new shape variables (form, circularity, and elongation) illustrate how far from a circle the shape deviates. Two additional variables to estimate size were created using the length and the width measurements: perimeter and area. These are two-dimensional shape and size variables that only describe the planar surface morphology.

The size of a grinding element allows us to compare surface dimensions between sub-features, features, and sites. Bedrock ground stone surfaces should vary in size depending on the type of grinding process (for example, food processing, toolmaking, hulling processes) and, in some cases, the material processed. Comparing size variables between BGS features may help us understand whether all activities on the different sites were the same or different and reflect a difference between adult workers or children mimicking adult workers. Lynch (2017a, b) discusses morphometrics and shape analysis in more detail. Perimeter measures the circumference of the grinding element using both length and width. For circular (elliptical) shapes, perimeter (P) is calculated using the formula to estimate the circumference of an ellipse: $P = \pi[3(a + b) - \sqrt{(3a + b)(a + b)}]$. Area (A) is calculated using the formula for the area of an ellipse: $A = \pi(ab)$.[3]

The shape of the grinding element informs a variety of behavioral processes. The morphology can indicate the movement of the milling strokes (push-pull reciprocal movements, circular rotary movements, or a combination of movements) and the size of the hand tool used during resource processing (elongated two-handed manos create flat, rectangular, and larger grinding surfaces; one-handed manos create more elongated or elliptical basins; and pestles create deep, circular, and u- or v-shaped basins). Circularity measures the deviation of the basic form from a perfect circle and is calculated as: square root of ((length times width)/(length squared)) or $\sqrt{((L \times W)/(L \times L))}$. Thus a circle = 1.0 and groove shapes closer to 0.4 or 0.5. Elongation calculated as *w*/*l* measures the stretch of a shape towards the lateral boundaries. Form, calculated as *l*/*w*, measures the stretch of the shape to the proximal and distal edges.

Lynch (2017a, b) developed an overview of the type grinding elements and features within the JE Canyon side canyon using the variables above. Lynch's analysis illustrates that there is a correlation between the length, width, and depth variables. Since these variables are the basis of the size variables, there was a positive relationship between these new variables. However, Lynch's analysis showed that the shape variables, although calculated with the same metrics, varied. Circularity and elongation variables are not significantly related to the length of the grinding element but are correlated with the width and the depth of the grinding element. The form variable is not significantly related to the length but rather negatively correlates to the width and depth. The perimeter variable is not related to either circularity or elongation variables but is related to the form variable. The area variable is significantly and positively correlated with both the circularity and elongation variables but is negatively correlated with the form variable.

These fundamental relationships allow us some flexibility to use either of the size variables (perimeter or area) in context with either the circularity and form variables or all three shape variables to describe the morphology of BGS feature elements. Shape analysis can get tricky fast, so the best variables provide good descriptions while keeping the math manageable. The variables permit us to visualize the shape using numbers to compare relatively easily within and between sites. For instance, form is a useful measure of shape because elongated groove-like elements—which have values between 4 and 6—are generally smaller. In contrast, circular to elliptical-shaped elements with values between 1.0 and 2.5 are generally larger. The exceptions to this trend are the small conical shapes with the lowest value for form (1.0) and relatively smaller surface areas. The important message here is to use the combination of the shape and size variables that provide the best description of the shapes observed on the site.

HYPOTHESIS TESTING

One of the benefits of using these shape and size calculations is determining whether there is enough difference between different BGS features to determine if site use varied. The sample at JE Canyon seems to be different from other areas along the Chaquaqua Plateau. Within a relatively small area, there are eleven sites with evidence of intensive grinding occurring at most of them, even those found in open-air contexts (Lynch 2017a, b). So, determining if the ground stone elements vary in terms of shape and size builds a comparative model that is useful across the Chaquaqua Plateau. We may not determine what was being ground, but we can say that grinding activities were similar or different, then try to isolate what might account for behaviors pattern.

The assumption underlying the shape analysis is that similar processing events for similar items (for example, food resources) should result in similar patterns, especially if the people doing the work were related or shared cultural, behavioral norms. Therefore, different grinding elements (either in morphological or quantity of grinding elements) deserve further investigation. The differences could be the result of demographic variation (age groups) or shifts in cultural behavior. The JE Canyon sample showed variations in BGS feature element morphology between the sites located at the head of the canyon, nearer to water resources, and those located along the canyon rim (Figure 3.2). Some of the more intensively used features were located far from water resources (Lynch 2017a, b). One of the more interesting results of the JE Canyon survey concerns the grinding elements found in open-air contexts. The four open-air sites were not as intensively utilized, but all had oval-shaped basins that exhibited shaping from a grinding process. Three were located near 5LA11446, on the upper rim overlooking the site, while the fourth was located near a large rock art panel (5LA11454), a few minutes' walk from the site. Lynch's (2017 a, b) analysis also showed a difference between the length, width, and depth of grinding elements between sub-features, features, and sites across the side canyon. While the sizes varied through the sites, their general form (oval-shaped) did not vary much between sites and features.

SUB-FEATURES

Sub-features are one way to analyze the relationship between grinding elements on a boulder or outcropping rock. In the JE Canyon sample, some of the grinding elements seemed to border one another in clusters. Some of the grinding elements were oriented in the same direction, while others were oriented perpendicularly or even overcut nearby basins. Lynch (2017a, b) grouped these closely related ground stone elements into sub-features to better understand how features may have been used.

The sub-feature classification is completed in the lab. Grinding elements that share a border are considered the nearest neighbor that may have been used in multi-step processing. Depending on where the border is shared, it might indicate moving the resources between basins, as indicated in the McCarthy et al. (1985) study, or they may have been used as staging areas, as True (1993) discussed. Additionally, there may be polished areas between grinding elements that could represent complex grinding processes. An example of such a complex grinding process requires grinding on one surface to achieve the desired flour consistency before moving the flour to another surface for additional processing. Description of Pueblo milling processes provides examples of multiple process-oriented constructed spaces for the Southwest (see chapter 8).

Descriptive illustrations of the relationship between grinding elements aids in the construction of meaningful and replicable analyses of the BGS features. A conceptual model of social space infers that workers would prefer to orient their bodies to maximize the ability to communicate with one another or supervise young children or other workers better. Orientation of the worker to see the spectacular canyon viewscapes may also be a consideration when looking at the grouping and orientation of BGS feature elements (Lynch et al. 2017). Clusters of BGS feature elements might illustrate social grinding spaces that incorporate adolescent grinders under the supervision of experts or mature family members (chapter 8).

DETERMINING TIME

Placing the bedrock ground stone features into regional chronology is problematic. Without direct ethnography, it is challenging for archaeologists to estimate just how long BGS features and the grinding elements on them may have been used before they were abandoned. Lynch (2017a, b) developed an approach to consider how much time may have been spent at a feature. This method involves estimating the amount of time needed to create and maintain individual elements. The method can be applied to estimate how long sub-features and features could have been used.

The method uses the depth measurement taken on each element in the field. The assumption is that the resource processing created the observed depth. The amount of use-wear from resource processing along with regular maintenance and retouching of the surface for efficiency continually changed the BGS feature element until it was no longer useful. While there is no experimental data on how long it takes to create the depth observed on BGS feature sites in southeastern Colorado, it is possible to use a proxy wear rate developed from Wright's (1993:350) experimental study. Combined with ethnoarchaeological data from Searcy (2011) that discusses average processing times for grinding daily meals for households and ethnographic evidence from the western United States, it is possible to develop highly generalized estimates of a time investment at BGS features and sites.

Exploring time investment at sites requires good depth measurements in the field. The formula to determine the minimum use rate is observed depth, in centimeters, divided by Wright's produced depth (.06 cm); the results are then multiplied by 53.5, the number of hours it took Wright to create the observed depth. In the case of 5LA11446, the average depth observed for all elements (N = 28) is 2.54 cm. Thus, one quick way to estimate the time invested in all 28 grinding elements is ((2.54 cm/.06cm) x 53.5 hours) x 28 the total number of elements. The result of the first part of the formula is 2,229, which is the average number of hours it took to create 2.54 cm of depth for one element. For all 28 elements, the time invested is 62,417 hours. This time estimate is refined below but provides a very general way of thinking about the time spent on a site.[4]

The next question to ask is how this translates into years. Was the site in use every day for a certain number of hours, or was it only used for a short time during a specific season? How long would workers be at the site if they used it every day? Would this differ if they only worked at the site during a particular season? At 5LA11446, there is no evidence of long-term habitation or intensive habitation on or close to the site. Evidence, on the contrary, points to seasonal use. In the absence of corn or other horticultural artifacts, it is safer to assume a hunter-gatherer mobility strategy in which the site was visited annually during certain seasons, defaulting to the pattern from the western United States. These visits were probably during the late spring and late summer when local food resources (for instance, pinon, juniper, oak, grasses) were fruiting. To calculate the total time of habitation in years, as represented by the depth, Searcy's estimate of 4.5 hours per day grinding for a single-family household (Searcy 2011) was used. The season was calculated at 91 days. Thus, in the example of 5LA11446, the depth took about 152 years to create (62,417 total hours/4.5 hours per day/91 estimated workdays per year). While very interesting, these are only estimates. Below we will look at how estimates change depending on how site BGS features are organized.

APPLYING ANALYTICAL METHODS TO 5LA11446

Site 5LA11446 is only one of eleven BGS feature sites in the small side-canyon surveyed, so it should be viewed as part of the system of sites (Lynch 2017a, b). There are 206 BGS feature elements in the JE Canyon sample; only 28 are at 5LA11446. Nonetheless, the site provides a good opportunity to apply the analytical methods introduced here to a single case and explore complications in BGS feature analysis.

All of the BGS features at 5LA11446 are within a single rockshelter that overlooks the Chacuaco Canyon floodplain. A dry-laid rock wall was constructed at some point in the rockshelter. The wall covers at least one BGS element; thus, we assume that the grinding elements were created before the wall. The ceiling height means that workers would have to sit or kneel to work. There are three sub-features (SF). Sub-feature A has four grinding elements, SF B has eight grinding elements, SF C has four elements. The remaining twelve elements are scattered throughout the shelter.

Each subfeature could represent a workspace for a single worker. However, the rockshelter workspace may have been "owned" by multiple workers from a single household or three separate households. Regardless, at least three individuals could work in the space simultaneously, either at the sub-features or at the other grinding elements. Thus, instead of thinking about the time investment at the site level, we should consider the investment at the sub-feature level as this will refine the use of depth measurements.

The four grinding elements in sub-feature A are morphologically oval-shaped basins characterized by circularity (C) of .81, form (F) of 1.56, elongation (E) of .65, perimeter (P) of 64 cm, and area (A) of 301 cm^2. The average depth of the elements is 2.1 cm. The average time investment is estimated as 7,490 hours (1,873 hours per element) over nine seasons.

Sub-feature B has eight elongated oval elements that are characterized by C = .83, F = 1.52, E =. 78, P = 68 cm, A = 391 cm^2. Although slightly larger than those in SF A they are still close to the average shape characteristics for the site (C = .80, F = 1.63, E = .64, P = 69 cm, A = 367 cm^2. The average depth for sub-feature B is 2.5 cm (deeper than SF A but close to the site average depth of 2.5 cm). The amount of time workers may have spent at sub-feature B is estimated as 17,833 total hours or twenty-two seasonal visits.

The four elements in sub-feature C are also elongated, oval shapes. The average shape is characterized as C = .75, F = 1.87, E = .56, P = 74 cm, A = 419 cm^2. These are, on average, more elongated and larger than the other sub-features and the site averages. The average sub-feature depth is 2 cm, so the estimated time spent working at the sub-feature is 7,490 hours over nine seasons, very similar to SF A.

The remaining elements are considered at the feature level. Feature 4 (the location of SF B and SF C) has seven individual grinding elements. The average depth is 2.3 cm. The average time investment is estimated as 14,356 hours over thirty-five seasons. Feature 5 (the location of SF A) has one additional element. The depth is 2.2 cm. Time investment is 1,962 hours over five seasons. Feature 7 has four individual grinding elements with an average depth of 3.1 cm. The estimated time investment is 11,057 hours over twenty-seven seasons.

While these estimates are curious, it is difficult to determine the actual site investment due to the number of possible scenarios. If the sub-features and individual elements are used to develop a site investment, rather than using all the elements on-site, the average time investment is less (59,831 hours over 106 annual visits) than the overall site average of 62,417 hours over 152 years. The reason is that people are working in groups at the same time, rather than as single individuals spread out over time. To complicate the situation, if the site was used only for ceremonial purposes throughout its life history, it might have been used for a much shorter time, say a week or two instead of the three months used in this example, but over many centuries instead of one or two as in this example. Additionally, the site may have been used for both ceremonial and quotidian purposes.

The overwhelming presence of the dominant hunting rock art panel suggests the site was used for ceremonial purposes or, at a minimum, reflected the symbolic relationship between hunting and grinding practices. Loendorf (2008:35) places the life history of the panel between A.D. 800 and A.D. 1000 with new figures added every few decades "when the panel was renewed to rekindle its potency." Thus, the hypothesized 200-year life history of the rock art panel and that of the BGS features are not too far apart if they were used together. An alternative hypothesis would be that the ceremonials associated with the hunting panel did not overlap with the grinding events coming either before or after, meaning the site's purpose changed through time.

CONCLUSION

This chapter covered topics from the best practices for manually recording BGS features in the field to methods for analysis and interpretation. The methods can be tailored to the research design and objectives. At the very minimum, the number of BGS features and grinding elements should be made in tandem with feature and site sketch maps. Field notes are extremely important to better understanding the regional prehistory. The methods offered for analysis and interpretation are heuristic tools for teasing out the site history but should be used judiciously.

Using a single site, 5LA11446, from the JE Canyon survey, the minimum data collected in the field can be used to extrapolate information about the continuity of BGS feature elements between the sites within the region. For instance, those open-air sites in the JE Canyon survey close to 5LA11446 are very different from the open-air sites that dominate the regional sample. These open-air sites have defined basins, are found in clusters of elements and have depth. The only amorphous shapes in the JE

Canyon survey are due to erosional processes that obscure the grinding basin. What can account for this difference? Are there other canyons with signatures similar to that of JE Canyon?

Finally, the case is made that by collecting reasonable accurate measurements, a regional model of grinding practice emerges. At the minimum, functional shapes vary and different elements were used together. Systems of grinding practice in bounded landscapes like those in JE Canyon help build a regional model. The life history of the BGS features can be explored by estimating the length of time they may have been used.

There is still much to learn about the nature and organization of BGS features in southeastern Colorado. We can be reasonably confident that some features and sub-features, especially those with clearly repeated morphological patterns, indicate complex, social grinding activities. However, we are still unsure how to determine whether individual grinding elements may have been used in quotidian or ceremonial activities. Three-dimensional recording methods provide clear, precise cross-sections of the grinding elements and an analytical tool for refining our understanding of BGS feature element morphology. We will explore the process of creating and using 3D models in the next chapter.

NOTES

1. A bounded landscape is one that appears to have a natural boundary within which sites are easily accessed. The types of sites may vary, however, because they are in such close proximity, it is tempting to assume that they were used by related people who reoccupy the site annually. The assumption has precedence in California land-tenure practices as well as in descendant Plains communities such as the Diné. Although the sites are related by ease of movement across the landscape, archaeologists still assign Smithsonian Numbers to artifact and feature rich areas that between 25 and 50 meters apart. The method was used while recording the sites in JE Canyon even though it is highly likely they are all related to one another.

2. The USGS Geographic Names Database (http://geonames.usgs.gov) lists four variants of the name: Chaquaquo, Chaquaqua, Chiquaqua, and Chacuaco. According to the USGS names decision card of 1910, Chaquaquo was used on the 1877 Wheeler's survey (USGS Geographic Names Information System). Chaquaqua Creek and Chaquaqua Canyon are the spellings on the 1890 to 1909 publications of the Mesa de Maya Sheets published by the USGS. In 1910 the Board on Geographic Names subsumed all variants under Chacuaco which is Spanish for clown; rough or course; and a small furnace. USGS quadrangles, beginning in 1922, began using Chacuaco Creek or Chacuaco Canyon, however as late as the 1930 census Chaquaqua is used for the local political precinct and Fenneman (1931) also continued to use the earlier variant.

3. In both formulae the *a* is half the length and *b* is half the width.

4. Although a stimulating thought experiment, directly extrapolating wear rates for Chacuaco bedrock features based on Wright's experiment is problematic for many reasons. McCarthy et al. (1985) propose that mortar surface depths were intentionally created for specific processes; if this scenario applies to bedrock features in JE Canyon, depth may be the function of a process that includes both creating the surface and processing food resources, not directly depicting time investment. A further concern is that we do not know the type of food resource being ground or pounded, although pollen analysis strongly suggests local resources like oak and pinon. Wright's experiment used corn. In addition, the difference in the rate of wear caused by a two-handed mano compared to a one-handed mano, even on the same type of sandstone, will likely be significant. The predominately oval shape of the basins found in the project's side canyon suggests the most likely candidate for causing the shapes observed was a one-handed mano.

5

Photogrammetry Models and Analysis

Metate Cave Case

Metate Cave (5LA6575) is a large rockshelter located on the Pinon Canyon Maneuver Site (PCMS) near Trinidad, Colorado (Figure 3.11). At some point in the past, local inhabitants of the region began to use the Dakota Sandstone boulder in the rockshelter for numerous purposes: food processing, symbolic transmission, toolmaking, and perhaps pottery production, among other tasks. Evidence for symbolic transmission is rock art pecked over the grinding elements as well as grinding elements imposed over rock art motifs. An independent rock art panel boulder stands watch over the site. More recently, visitors to the rockshelter engraved their names over many of the grinding basins. The site is unique in the local landscape because of the number of BGS feature elements. Eager to place the shelter's activities into the broader archaeological context, the main question Pinon Canyon archaeologists have been trying to answer, since finding the site, is *how* to best understand what human behavior occurred here.

Although the site was "discovered" by archaeologists during a cultural resource survey of Welsh Canyon (Loendorf and Loendorf 1999: 35–37; Schiavitti et al. 2001), area residents were familiar with the rockshelter. Unlike the JE Canyon survey discussed in chapter 4, where the BGS feature sites occur throughout the canyon, we have yet to locate any closely associated BGS feature sites near Metate Cave. The site is on the top of a ridge overlooking an intermittent creek (Figures 5.1, 5.2). The large rockshelter is easily accessed from the creek bottom or the rimrock above and offers abundant shelter, although the shelter temperatures during the summer are exceedingly uncomfortable.

Several distinct occupations occurred, based on the BGS features; however, there are no absolute dates for the shelter. The first impression one has upon entering the rockshelter is the number of tool grooves and the almost overwhelming number of grinding elements. The second impression takes in the names carved by locals in the early twentieth century onto the BGS features, sometimes directly within the grinding elements. Most of the twentieth century names date from 1900 to 1920. One name is from the 1980s.

Even the casual observer gets an impression that the grinding elements were probably used before the shelter was employed to make tools before the rock art was created because the superposition of elements is unmistakable. It is only upon examination that the sequences of occupation begin to unfold. Pecked honeycomb-like rock art motifs are *under,* and geometric motifs are pecked *over* some grinding elements. Tool grooves are ground *over* some ground stone basins but also through other tool grooves. The tool groove boulder seems to have been displaced from a horizontal position to a vertical one, probably due to erosion under the boulders.

Several hypotheses have been proposed for the rockshelter. One hypothesis suggests the site was a large food-processing area for incipient horticultural practices, yet we have no evidence of domesticated plants directly associated with the rockshelter. Another hypothesis proposes that the shelter was used for baking food during the summer months when the heat inside the shelter reached as high as 110° (Schiavitti et al. 2001:76). A third hypothesis considers that the rockshelter might be related to an Apishapa structure located in the vicinity, causing some archaeologists, myself included, to wonder how the BGS features relate to the Apishapa culture. If some of these features date to the Late Archaic (circa 2,000 years ago), and if the Apishapa cultural expression (circa 900 to 500 years ago) is local, how did BGS features change as semi-sedentary horticultural practices developed? If BGS features are entangled with the development of social identity and the maintenance of avenues for the social reproduction of knowledge (probably through labor identity), the presence of rock art (particularly the types of rock art motifs) might help understand these past social processes.

As with most promising science, any or all of these hypotheses could be plausible, but they remain speculation about possible past behavior without a means to test them. This chapter explores how 3D visualization can help us refine our understanding of archaeological features that resist traditional testing strategies. Since this is a relatively new set of methods, time will be spent discussing the planning of data collection in the field and the lab processing. Close-range photogrammetry (CRP) provides accurate, descriptive imagery (site maps and feature maps) and highly accurate datasets that can be used for further analysis.

SITE ENVIRONMENT

Metate Cave (5LA6575) is situated within a gently sloping drainage area (Figure 5.1). The rockshelter has been created by the steady erosion of the underlying sediments and bedrock. Large segments of the bedrock broke apart from the parent Dakota sandstone formation at some point in the past, creating a deeply recessed sheltered space with a bedrock floor. Erosional processes are still active within the rockshelter, as evident by the crumbling bedrock blocks slowly eroding into a small gulch-like crevice directly beneath the rockshelter opening.

Water movement appears to be the main culprit of erosion, as it is undercutting the softer material beneath the prominent boulders. These have been and continue to be weakened, causing the larger "capstones" to shift and crumble. Large cracks predict how the next boulders will separate. Existing shelf-like features and open spaces beneath the BGS features illustrate where gravity might place them. Wind erosion must also be a factor because the designs on the free-standing rock art feature (Feature 5) are significantly degraded. One more potentially destructive culprit should be mentioned. A large packrat midden is being constructed under the main BGS feature. The visible aspects of the midden suggest a recent use of not more than fifty years; however, deeper, older deposits may be hidden under the boulders. Nonetheless, packrats are actively contributing to the gradual changes within the shelter.

How this landscape looked to the earliest inhabitants who used the site for grinding, tool making, or ceremony is impossible to speculate on without more robust dating methods, so we have to content ourselves with understanding the modern vegetation and climate that may have been similar in the past. While colder during the fall and winter months, summer witnesses a spike in temperature within the shelter because of the western orientation. Several boulders from the rimrock lie scattered in the small gulch at the mouth of the rockshelter. During rainy seasons, the shelter may hold pockets of water, but it remains relatively dry. A catchment in the small gulch at the shelter base may provide a semi-stable water supply for activities within the shelter.

The dynamic, potentially destructive, erosional aspects of the rockshelter made it imperative that we develop a 3D model of the shelter. Thus in 2011, at the end of the JE Canyon research project, the author visited the site with Larry Loendorf and Mark Owens (former PCMS archaeologist) to plan for a virtual site preservation plan. The author returned in 2016 with Jennifer Kolise (PCMS archaeologist) and close-range photogrammetry experts Tom Noble and Neffra Matthews (both retired BLM). We had one full day to capture 3D data and fully record the features. Due to the time crunch, Jennifer Kolise surveyed the surrounding area for potentially related archaeological features. At the same time, rockshelter recorders divided us into two teams: a manual recording and mapping team and a 3D modeling team, based on our relative strengths in the respective recording methodologies.

When archaeologists recorded the site in the past, they had developed a series of feature maps for the shelter and rock art drawings (Figure 5.3; other sketch maps were published in Schiavitti et al. 2001). After confirming their accuracy, these sketch maps were used as base maps during the 2016 recording to save time. The feature numbers assigned in the original recording were maintained, but we developed a new identification system BGS elements because only a few had been explicitly identified on the original sketch maps. As with the JE Canyon methods, blue masking tape was used to number the features and elements. Photographs were taken of the site identification labels to check the field recording processes. The original sketch maps were then enlarged and edited in the field to track the new element identification. Thus, we created two backup systems to ensure replicability of the element measurements. The sub-features were created in the lab after we completed the 3D models.

DEFINITIONS

Before continuing to the recording methods, we should review the exact terminology to be used, the definitions applied to them, and assumptions inherent in the language. These definitions have been developed through time but rely mainly on recording processes used by the PCMS archaeologists and OAHP guidelines. They may differ somewhat from JE Canyon research, published elsewhere, because they combine established recording terminology with an evolving set of practices as new sites are discovered and newer methods of analysis developed.

- Feature: A feature is an object or objects fixed in place. In describing bedrock ground stone, the term is applied to an area of bedrock material (either boulders or exposed surface areas of caprock) that contains a set of one or more elements (grinding basins, tool grooves, or rock art panels). The term *artifact* is not used for bedrock grinding elements because they are, by nature and design, non-portable. One assumption of this category as explicitly applied to BGS features is that they represent self-contained workspaces where one or more people could, and did, process resources together. There may be multiple features within a given locality or site, like a rockshelter. A corollary assumption is that distinct features represent *different* workspaces—either different people or different processes. Depending on the size of the space, they may or may not have been used simultaneously.

- Process: A process refers to the type of work performed at a feature. It might be grinding food resources, creating tools through reciprocal grinding motion, or creating temper for pottery through pounding and grinding action. In the case of food or pottery, the grinding process includes another tool-stone (mano, pestle, or pounder) to reduce the materials to an appropriate (as determined by cultural preferences) powdered substance. In tool creation, the observed archaeological element was created by grinding on the bedrock surface (Adams 2002; Feyhl 1980; Stoffle 1984). An example of this would be a piece of bone ground against the bedrock surface to create a bone awl or needle. The observed tool groove was created by attrition. The process might leave a detectable residue (bone) in the surrounding sediments. The *process* of creating rock art motifs in Metate Cave shelter consists of two main types: pecking and incision. Pecking uses other stones or materials to physically remove bits of the parent rock to create a design motif. Incised motifs could have been created with any sharpened, hard material (bone, stone, or even pin knives in a historical context).
- Element: Bedrock ground stone features may represent many types of human activity or processes that leave behind residual products of that activity. Element is used to characterize the individual parts that make up a bedrock feature. In other BGS recordings, the term field specimen (FS) has been used (Lynch 2017a, b). However, since FS usually refers to collectible artifacts like projectile points or other portable tools, this term creates confusion when applied to BGS features. Although the term "element" has been used for a while in rock art recording, BGS feature recording uses numbers instead of letters. At Metate Cave, the main elements include: 1) grinding basins (also just called basins) with shape sub-divisions characterized as oval or circular; 2) tool grooves characterized as complete or incomplete; and 3) rock art motifs characterized as pecked linear, pecked circular, pecked representational, incised historical names and dates. Elements should always be referred to in association with their feature and sub-feature in site recording and reports.

RECORDING METHODS

The limited time available to access Metate Cave (5LA6575) necessitated developing a new approach to accomplish the main goals; providing replicable measurements of the features on the features, capturing stereo-overlapping photographs for 3D modeling, ascertaining if there were other sites nearby. All previous recordings and cultural resource reports were accessed through the PCMS and OAHP databases in preparation for the fieldwork. The only published materials are cultural resource management reports; otherwise, the material is field paperwork. However, some of the original site recorders were still working at the PCMS during site visits, so they provided additional information to help plan the recording procedures. The original field maps were highly accurate in terms of site and feature details. Although the field recordings were sound, there was no way to reproduce the measurements discussed in the field reports, so the grinding elements (grooves and basins) were re-labeled and re-measured. We did not re-label rock art, although the panels were photographed for 3D models.

The recording team was divided into three units: one recording the elements, a second collecting the photogrammetric data, and the last to survey the surrounding area for additional features or sites. The manual recording process quickly assessed the previous maps' viability and then began new feature recording forms. We used blue tape to label elements on features as we progressed so that the photogrammetry team could work in areas without tape. They gathered test photos of the features to provide additional backup for element identification in the lab. We used these to develop new feature planviews for the sites with element identification (publication forthcoming). Data gathered in the field were entered into an excel spreadsheet. Shape analysis and morphology of the BGS feature elements were analyzed to explore patterns within the rockshelter but also to compare with the morphometric data at JE Canyon Ranch. Simple descriptive and exploratory statistics were developed using SPSS.

Noble and Matthews developed a 3D model of the entire rockshelter (Figure 5.3). This model is available for visualization purposes and has been used through the Shell 3D Visualization Center to teach introduction to archaeology and highlight CAV technology during open houses. Photographic data were archived at the University of Wyoming and used to develop the feature maps presented below.

Photogrammetry processes used Metashape by Agisoft (www.agisoft.com) to create 3D field maps. The processes followed Noble and Matthews (2016, 2018). Archival photo sets of the raw and jpeg files were created following Noble and Matthews (2016) and CHI (http://culturalheritageimaging.org) recommendations. Using Adobe Bridge, the files were renamed into two folders: for camera raw and jpeg files. Raw data images were changed to digital negatives (.dng) files. Digital negative files were used to build the models in Metashape using Noble and Matthews (2018).

The purpose here is to present methods to describe and document any feature with 3D visualization methods that can then be made accessible to other researchers and illustrate some of this process's results. At 5LA6575, there are three types of features: the rockshelter, the BGS features, and the rock art panels. In tandem with this process is developing the means to analyze the descriptive data (measurements of the elements).

RESULTS

There are 129 BGS feature elements at 5LA6575 distributed across four features (Figure 5.3; Table 5.1). The original field recording broadly referred to bedrock surfaces on Feature 5; however, there are two distinct boulders: one with rock art, the other with grinding basins. In addition, there are small chunks of boulders with ground stone surfaces laying on the floor of the rockshelter. Although we could have fitted them back onto the parent boulders, they, along with those on Feature 5, were placed in a separate category of ground stone isolates. Ideally, the latter warrant the creation of a sixth feature. For parsimonious reasons, a free-standing boulder near Feature 2 was recorded with that feature as a sub-feature.

The nature of archaeological features, particularly those related to BGS features, is discussed later in the chapter. However, perhaps like rock art panels, a different category needs to be developed that explicitly considers site function. For instance, space seems to dictate the number of BGS feature elements. Simply put, the more space available to create work surfaces, the more elements we find. Feature 2 (Figure 5.4) is quite prominent, so understandably, there are more grinding elements. This assumption seems intuitive, but other, as yet unknown actors should be considered if these are socialized work environments. This inquiry line also questions how many people could be accommodated at the individual features or within the rockshelter. If we only use the recording team as a real-world experiment, two teams of two people (with their equipment) worked quite comfortably at all of the features. The only constraints were the ceiling height, which becomes more restraining towards the back of the shelter and at the edges of the boulders themselves.

Table 5.1. Summary Data of All Features and Elements at the Metate Cave Site

Feature	*Elements (N)*	*%*	*Area (cm2) average*	*Ceiling height average*
1	41	31.8	116	Na
2	51	39.5	320	1.12
3	2	1.6	556	1.3
4	20	15.5	569	1.93
Isolates	15	11.6	251	2
Total	**129**	**100.0**	**250.7**	**1.45**

Sub-features were designated for those elements that are in close association or touching one another. The category was developed to better understand the relationship between elements within features and on the sites. It is also a valuable category for comparison with other sites in the region. There are ten sub-features on the site. The largest sub-features A and B are on Feature 1 (Table 5.2). Sub-features C through I are on Feature 2, although sub-feature I could be considered a different feature. Sub-feature J is on Feature 4. The category "not assigned" means that the element was too far from clusters to assume association. It also contains those elements that were found on the floor of the rockshelter. The most interesting result is the relatively small clusters of elements, except for those on Feature 1. These results suggest that sub-feature B (made up of bigger grinding elements, as opposed to the grooves in sub-feature A) may contain two sub-features (Figure 5.5). The space on this feature was highly valued.

There may also be problems identifying grinding sub-features when other types of work have usurped the original workspaces. For instance, the types of elements on Feature 1, sub-feature A are predominately grooves, while those on Feature 1, sub-feature B are predominately deep oval basins. The digital elevation model (DEM) of Feature 1 shows an apparent superposition of grooves over both basins and other tool grooves (Figure 5.5). The superposition of different types of ground stone elements makes sub-feature designation a bit of an art. In rethinking sub-feature distribution, it is possible to lump all like elements (Figure 5.6 a) or split the sub-features into smaller clusters (Figure 5.6 b).

The average size of the elements in each of the sub-features, on Feature 1, allows us to compare relative workspace. We used two of six morphological characteristics from the previous chapter, the perimeter and area. Perimeter measures the planar rim, while area measures the planar surface area. The planar surfaces are the final surface shapes that remain after the workers abandoned the features. One of the reasons we have so many indeterminate measurements and shapes is the superpositioning of grooves and rock art over the basins. While the boundaries of the elements are easy to see, it is difficult to determine if the effect is due to intentional design.

The critical issue is how to visualize the BGS features to answer specific questions. This chapter aims to illustrate the methods that best describe the BGS feature at 5LA6575. Table 5.2 used the most straightforward breakdown of elements into two sub-features. The process illustrates a difference between Feature 1 and the other features at the site thus, skewing the interpretation of feature and sub-feature use. However, the average size of elements in sub-feature B is higher than those in sub-feature A, something we expect to see given that most of the elements in sub-feature A are grooves.

Table 5.2. Summary of the Sub-Features at 5LA6575

Sub-feature	*Elements N*	*%*	*Perimeter (cm)*	*Area (cm)*	*Circularity (cm^2)*	*Form (cm^2)*	*Elongation (cm^2)*
A	25	19.38	33.01	75.56	0.55	4.35	0.33
B	12	9.30	63.17	271.93	0.69	2.74	0.50
C	4	3.10	64.06	318.08	0.77	1.82	0.61
D	3	2.33	74.37	402.11	0.77	1.76	0.60
E	6	4.65	55.82	240.02	0.61	4.58	0.44
F	7	5.43	87.62	567.47	0.81	1.56	0.65
G	7	5.43	59.50	280.18	0.67	5.17	0.50
H	6	4.65	81.76	474.76	0.76	1.75	0.57
I	3	2.33	62.57	274.88	0.76	1.79	0.57
J	6	4.65	83.95	560.23	0.85	1.45	0.72
not assigned	50	38.76	56.61	262.12	0.60	5.67	0.42
Total	**129.00**	**100.00**	**55.17**	**250.76**	**0.65**	**4.00**	**0.46**

Feature 1 probably has more than two sub-features. Indeed, because the grooves overcut the basins, there were at a minimum two distinct types of grinding on this feature. The pattern of many grinding elements in sub-features is repeated on Feature 2, sub-features F, H, and I. Further investigation, comparing digital elevation models between the sub-features at 5LA6575, helps refine the use scenarios for the site.

A BRIEF NOTE ABOUT THE ROCK ART AND GRINDING ELEMENTS

There are two main prehistoric rock art panels at 5LA6575. One is on a large roof fall boulder that faces the north (Feature 5 but referred to as Panel 5 in 1995). The position of the BGS feature elements near the base of the rock art panel suggests at least two plausible scenarios: the elements were used before the boulder cleaved from the ceiling (thus before the rock art panel was created), or they were used in conjunction with the panel. The motif is a linear pecked design with concentric circles. The design is similar to one found in JE Canyon, although the positioning of the motifs is different. There is at least one horned-quadruped.

Digital elevation models were developed of the panel, which is highly eroded (Figure 5.7). Although the panel was recorded in detail and analyzed by Larry Loendorf, it seems to be under heavy erosional stress. Incorporating the panel into the 3D rockshelter model means that it is somewhat preserved, at least the condition as of the 2016 recording. The second rock art panel is on Feature 2. The most obvious rock art motif is a linear pecked design with circles pecked at the ends of intersecting lines (Figure 5.7). The motif has been pecked over the grinding basins, indicating that the pecked motif post-dates the BGS feature. The line runs in an east-west direction similar to Feature 5. 3D models helped identify additional rock art motifs *under* grinding basins (Figure 5.8). These are two main series of honeycomb-like designs moving eastward from the edge of the parent boulder. Pecked designs were also noted under other grinding elements on BGS Feature 2 that appear to be tracks obscured by grinding in the basin.

DISCUSSION

During the period of occupation, the shelter seems to have been used intensively and regularly. However, it also appears that the rockshelter drops into obscurity for long periods. For instance, the twentieth century names cover only the periods 1900 to 1920 with sixty years until someone left their mark behind on the rockface during the early years of the PCMS. Considering other rock art localities in the region, visitors tend to leave a record of their sojourns through the area. Thus, the site seems to undergo periods of popularity punctuated by those of indifference until its rediscovery by archaeologists in the 1990s.

REGIONAL COMPARISON

Although we have some metric data from OAHP other sites across the region, the data are not generally related to specific elements. While these data types are acceptable for broad shape descriptions, with the caveat that they are not replicable, they are not useful for specific one-to-one comparisons with 5LA6575 data. Aside from Metate Cave data, there is only one other data set, at the moment, that can be replicated: the JE Canyon survey.

A simple t-test between the morphological characteristics, size (perimeter and area), depth, and shapes (circularity, form, and elongation), rejects the hypothesis that the shapes of the elements are similar. There are problems with this comparison. The main issue is that no site in JE Canyon has the number of tool grooves exhibited at 5LA6575; although 5LA11451 has a panel of tool grooves near a BGS feature, it was considered a separate workspace, so deleted from the BGS feature data.[1]

The Metate Cave sample is quite large with 139 BGS feature elements compared to the total JE Canyon sample of 209. The most extensive BGS feature at JE Canyon (5LA11453) has 68 grinding elements, only 35 of which contribute meaningful data due to erosion of some grinding elements. So 5LA6575 is quite a unique site for the region, although it has more in common with the JE Canyon sample than any other BGS feature sites in the region for which we have information. Figure 5.9 illustrates the differences in the morphological characteristics between the two localities.

The size (area and perimeter) of elements at 5LA6575 fall between the elements at 5LA11453, the most intensely used feature in the JE Canyon sample, and 5LA12788, one of the open-air sites above 5LA11446. The grinding elements are smaller, but not by much, than those at 5LA11446, discussed in chapter 4. The size variation relates to the presence of so many tool grooves in the measurements from 5LA6575.

A difference between the two samples is the categories of shape impressions on the recording forms. JE Canyon categories included terms like D-shape and rectangular for grinding elements that are much larger than any found at 5LA6575. In contrast, Metate Cave observations had more flexible categories for smaller gouges and pecked cupule-like marks found throughout the features.

While the Metate Cave grinding elements seem to resemble those in JE Canyon, in terms of the large quantity of BGS feature elements nestled within a rockshelter, a closer look at the shapes, sizes, and depths illustrates that the localities are quite different. One of the key differences is the layering of the rock art with the grinding elements. Discovering what these differences mean in terms of past landscape use and social identity will require different types of analyses.

CONCLUSION

This chapter illustrates that manual and digital recording methods can be completed quickly with proper training and equipment. In the case present here, we had a good foundational site map to guide data collection. As illustrated in the JE Canyon case, this is a rare occurrence. More often than not, surveyors will be faced with deciding how much time to invest in recording the BGS feature since so few have been thoroughly documented.

The take-home message is that even the most basic field map and photos can provide good descriptions of these features, thus helping to include them in regional archaeology. Metate Cave features are significantly different from those recorded at JE Canyon. At JE Canyon, there are more than 200 elements across eleven sites spread throughout a side canyon—all easily within walking distance of water and lush resources. On the other hand, Metate Cave seems relatively isolated from other BGS features but has more than half the elements. Like JE Canyon, there are ground stone elements with a distinct oval (elliptical) shape and numerous tool grooves.

The rock art in both samples suggests a symbolic link between the grinding spaces and hunting activities. Loendorf (2008) noted this pattern in JE Canyon. A particular element of interest is the concentric circle motif in both localities and quadrupedal motifs directly associated with the BGS features. The presence of symbolic representation with grinding elements is perhaps a clear indication that grinding activities were intricately linked with social identity and the social reproduction of knowledge.

The Metate Cave case illustrates that rockshelter sites played an essential role in the long-term human landscapes of the regions. Because the features are difficult to date thus hard to place in chronological sequences, superposition of elements coupled with seriation and comparative morphometric analysis are the most important tools we have to determine a relative chronology for BGS features. In both JE Canyon (5LA11453 and 5LA11451) and Metate Cave (5LA6575), abstract designs were pecked over grinding elements. At a minimum, the situation indicates that the grinding elements were abandoned since they were no longer used for grinding after the rock art was created on the surface. Whether it points to a ritual "killing" of the elements after the "owner" died, or a shift in the symbolic use of the space is impossible to elucidate at the present moment. However, at Metate Cave, we also have ground stone elements created over rock art panels.

The significance of this change in past behavior is not lost on most archaeologists in the region. Were there time intervals during which the importance of grinding work surpassed that of hunting? What changes might have affected the shift in worldview and the ceremonies that accompanied them? How can we use the superposition of rock art with grinding elements to better understand regional chronology? These questions cannot be answered today. However, with additional excavation near the rockshelters and a better recording of BGS features regional, especially using 3D technology, we will eventually answer these questions and others relating to sustainable landscape practices.

This chapter and the previous one discussed field studies with highly trained experts with technology but limited time (Metate Cave) and trained experts with limited technology but unlimited time (JE Canyon). Neither of these cases is the norm for regional cultural resource surveys. The next chapter illustrates how even a quick survey can generate high-quality data for BGS features analysis. It also synthesizes what can be done with manual methods and digital recording.

NOTE

1. The obvious relationship of BGS features and tool grooves, based on these two sites, means that a new method for comparison and site evaluation needs to be developed. 3D models of the tool grooves at 5LA11451 are currently being developed from photographs taken during the 2011 fieldwork.

Figure 1.1. E.B. Renaud and Janet Chatin during fieldwork along the Cucharas Drainage. Image used by permission of the Janet Chatin Collection, Huerfano County Heritage Society.

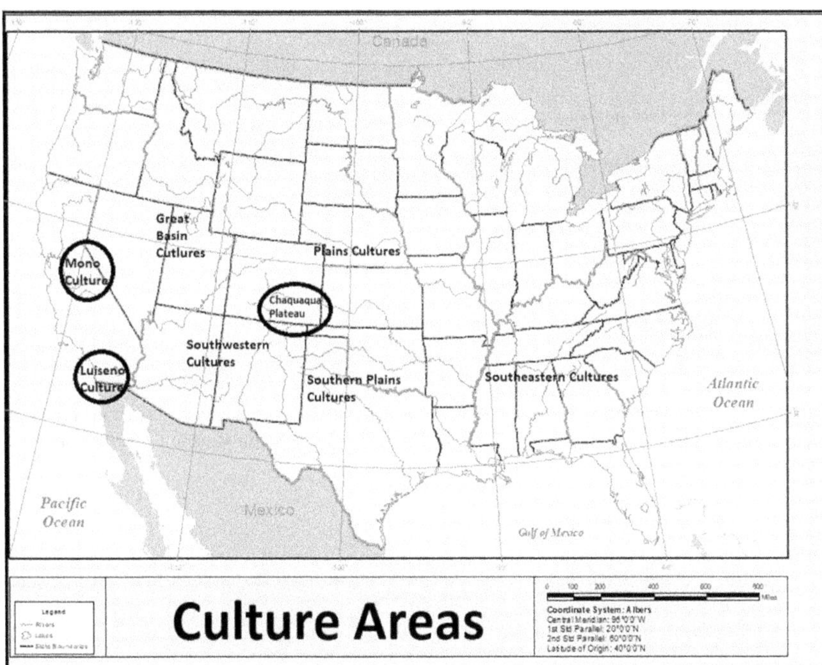

Figure 1.2. The Chaquaqua Plateau and other cultural areas discussed in the text. The Chaquaqua Plateau is sometimes included in the Central Plains or the Southern Plains culture areas (Wedel 1974:81, 130); however, it is more appropriately included with other High Plains cultures (Scheiber and Clark 2008:4; Wedel, 1963). Map Credit: Elizabeth Lynch.

Figure 1.3. The Trinchera bedrock ground stone feature, circa 1949. Photo by Herbert Dick. Louden-Henritze Archaeology Museum at Trinidad State Junior College.

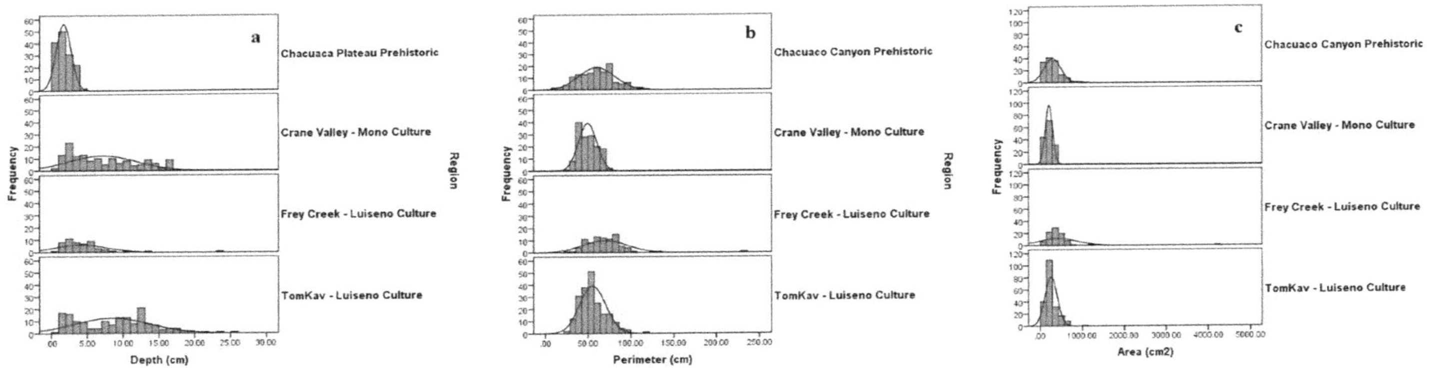

Figure 2.1. Histogram showing the distribution of grinding surface sizes between the Western Mono, Luiseño, and Chacuaco Prehistoric (JE Canyon Ranch Survey) cultural samples: a) depth in centimeters; b) perimeter in centimeters; and c) area centimeters squared. Created by Elizabeth Lynch.

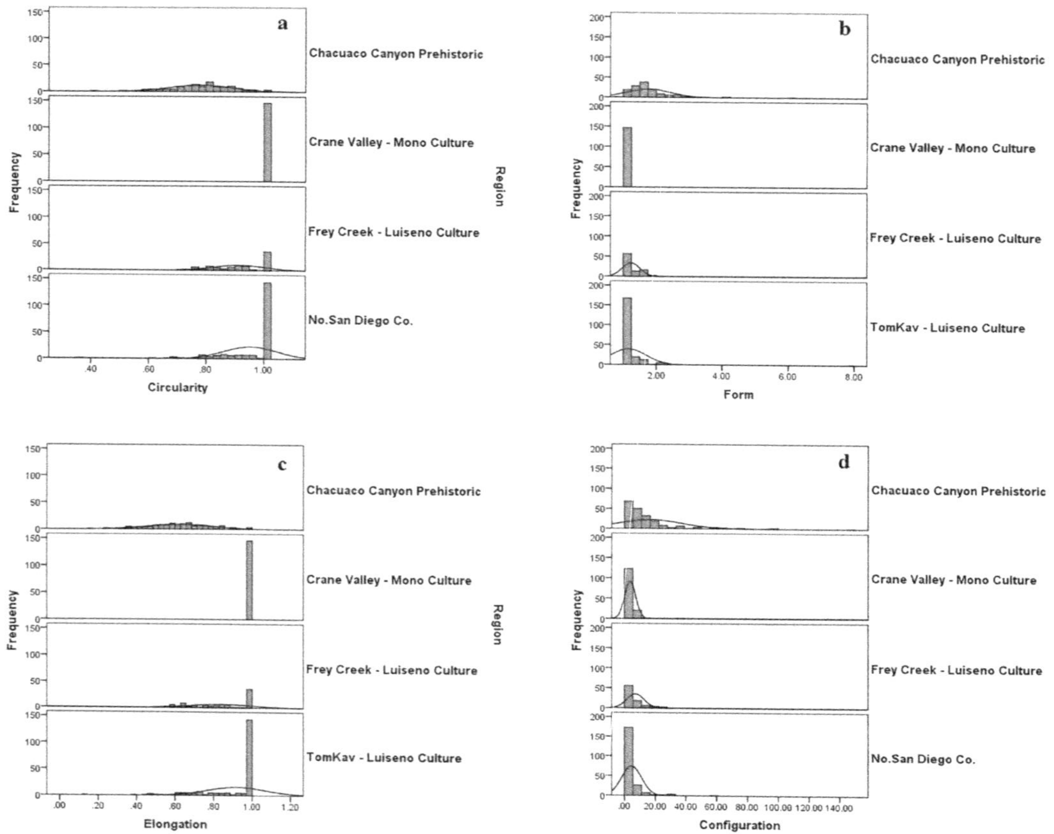

Figure 2.2. Histogram showing the variation in shape variables for each of the regions sampled: a) circularity; b) form; c) elongation; and d) configuration illustrate the differences between the regions. The Chaquaqua Plateau grinding elements, although somewhat similar to the Luiseño cultural expression, do not have the circular mortars characteristic of California bedrock ground stone features. Created by Elizabeth Lynch.

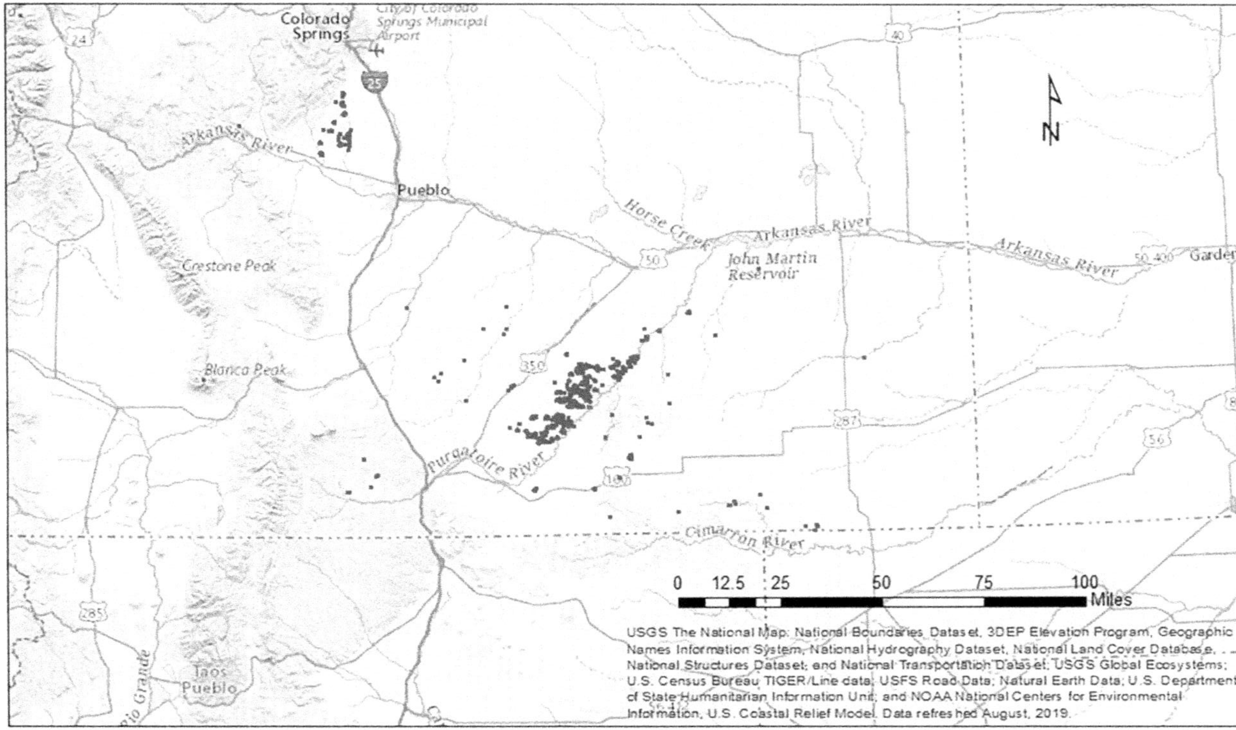

Figure 3.1. Map illustrating the distribution of BGS feature sites on the Chaquaqua Plateau. GIS data obtained from the OAHP, Denver, Colorado. Map Credit: Elizabeth Lynch.

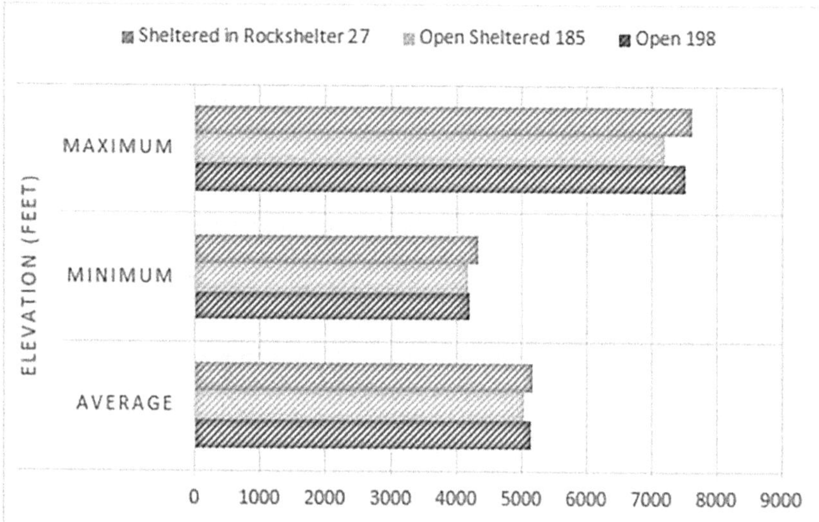

Figure 3.2. BGS feature site elevations based on landscape context of the three main viewshed descriptions. There are 198 open sites, 185 open sheltered sites, and 27 sites with BGS features in a rockshelter. There is little or no difference between the viewshed and the elevation of a site. Open sites and open sheltered sites are found across the range of elevations, as are those sites with clustered grinding features found within rockshelters. Created by Elizabeth Lynch.

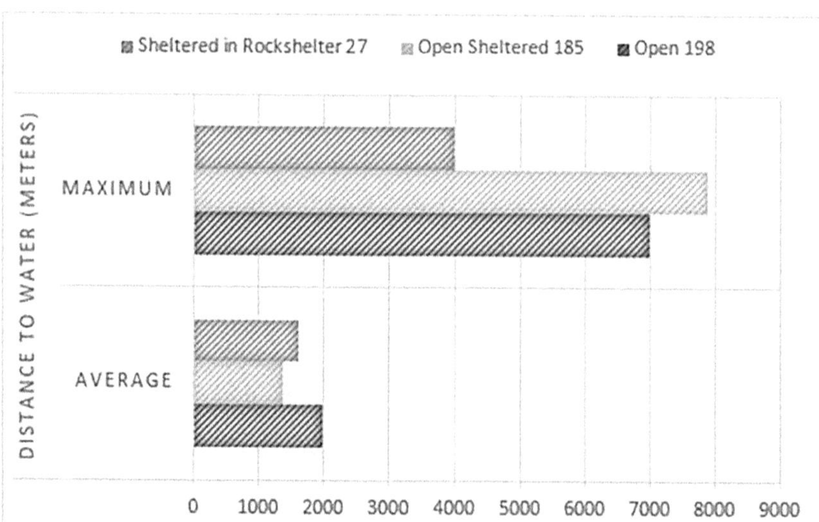

Figure 3.3. Sites with BGS features located within a rockshelter are closer to permanent water sources than those located in open-air locations. However, some sites with shelter are found nearer to water sources. The maximum distance to water for these heavily used sites is less than any other BGS feature site. Created by Elizabeth Lynch.

Figure 3.4. Map of related sites distributed within a single side canyon. The BGS features are found within rockshelters along the canyon rim and deep within the side canyon. However, note that 5LA11453, a site with the most extensively used BGS feature, is located further from permanent water. Created by Elizabeth Lynch.

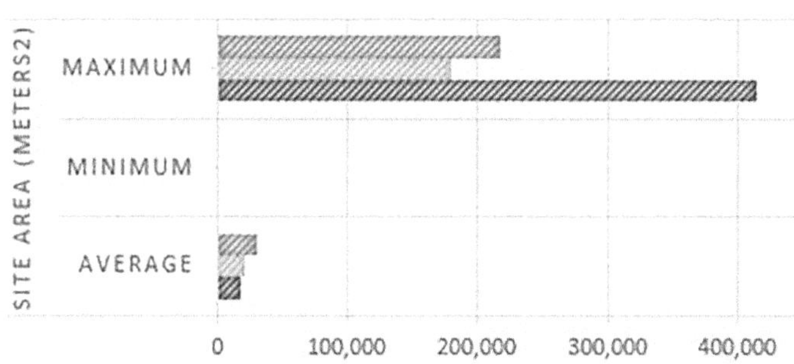

	Site Area (meters2)		
	Average	Minimum	Maximum
Sheltered in Rockshelter 27	31,151	23	218,400
Open Sheltered 185	21,790	20	180,313
Open 198	18,135	1	414,185

Figure 3.5. Sites with BGS features within a rockshelter (N = 27) are on average larger than either open-air but relatively sheltered locations (N = 185) or fully open-air locations (N = 198). Created by Elizabeth Lynch.

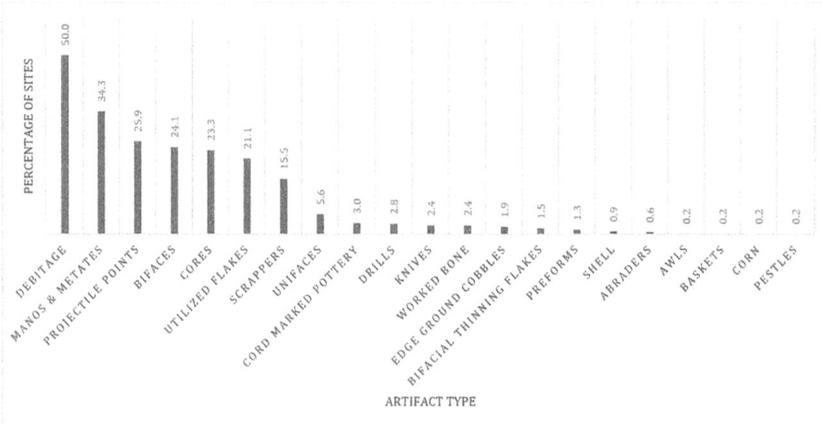

Figure 3.6. Sites with BGS features within a rockshelter (N = 27) are on average larger than either open-air but relatively sheltered locations (N = 185) or fully open-air locations (N = 198). Created by Elizabeth Lynch.

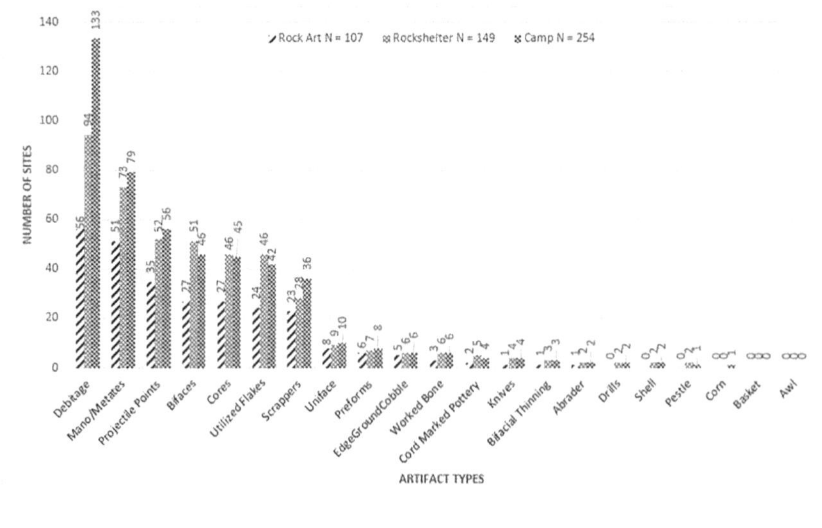

Figure 3.7. Artifacts present on BGS feature sites classified as rock art (N = 107) within or near a rockshelter (N = 149) or a camp (N=254). Created by Elizabeth Lynch.

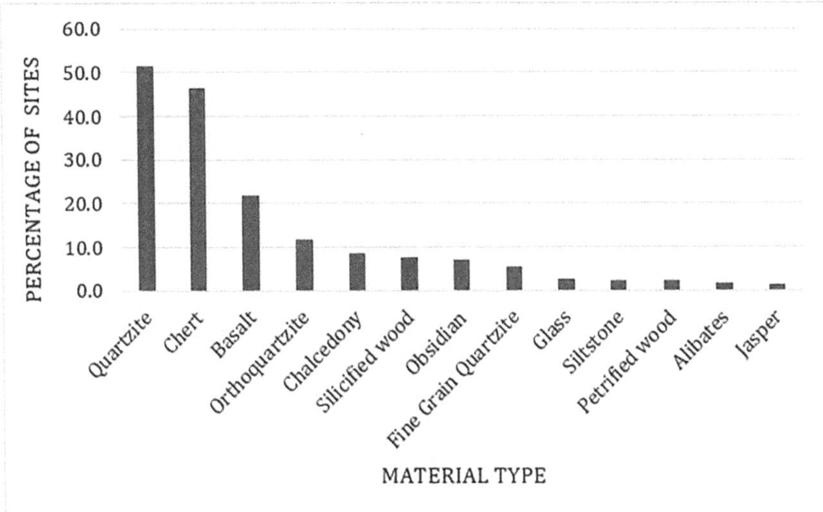

Figure 3.8. Percentage of BGS feature sites with different types of material. Not all sites have lithic materials. The absence may be a recording error, or the feature may be genuinely isolated. Created by Elizabeth Lynch.

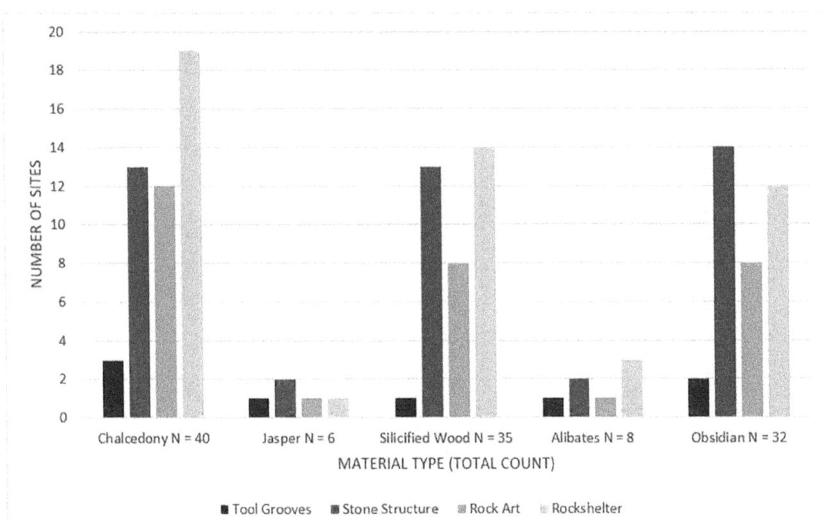

Figure 3.9. Number of sites with BGS Features that have non-local materials based on the description of the archaeology on site. Rockshelters near or harboring BGS features are dominant in all categories except those with stone structures. Created by Elizabeth Lynch.

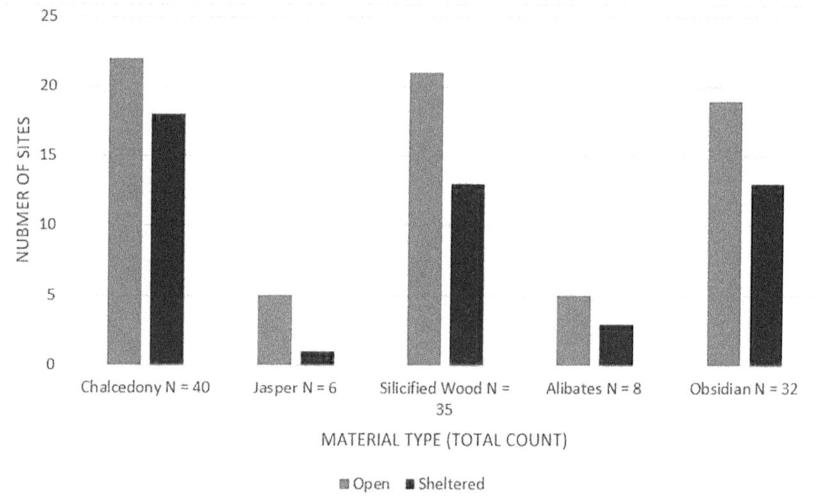

Figure 3.10. The number of sites with BGS Features with non-local material types, based on the viewshed description. Created by Elizabeth Lynch.

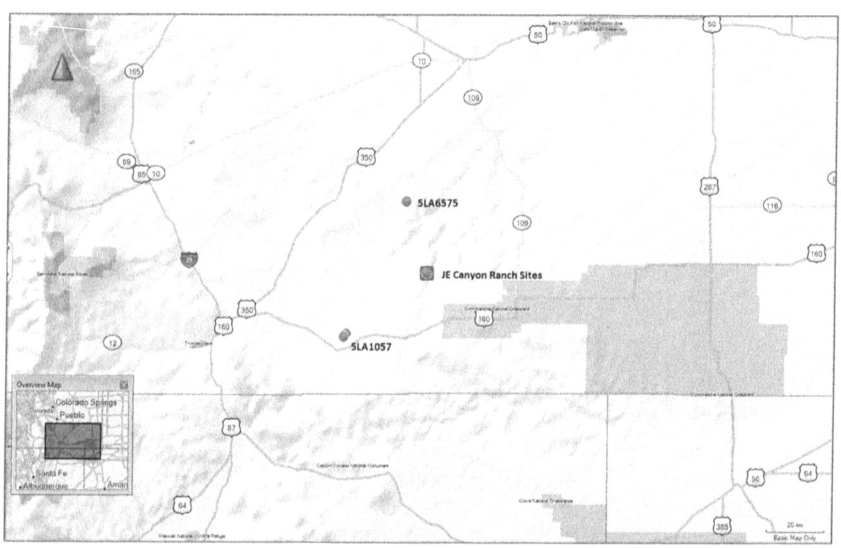

Figure 3.11. Map of the location of Metate Cave, Trinchera Cave, and the JE Canyon survey. The sites are rockshelters with the typical BGS boulder with clusters of oval or round-shaped grinding surfaces. The JE Canyon Ranch survey includes 11 sites. 5LA6575 is a single site located on the PCMS. 5AL1057 is the rockshelter associated with the Trinchera Cave group of sites. Created by Elizabeth Lynch.

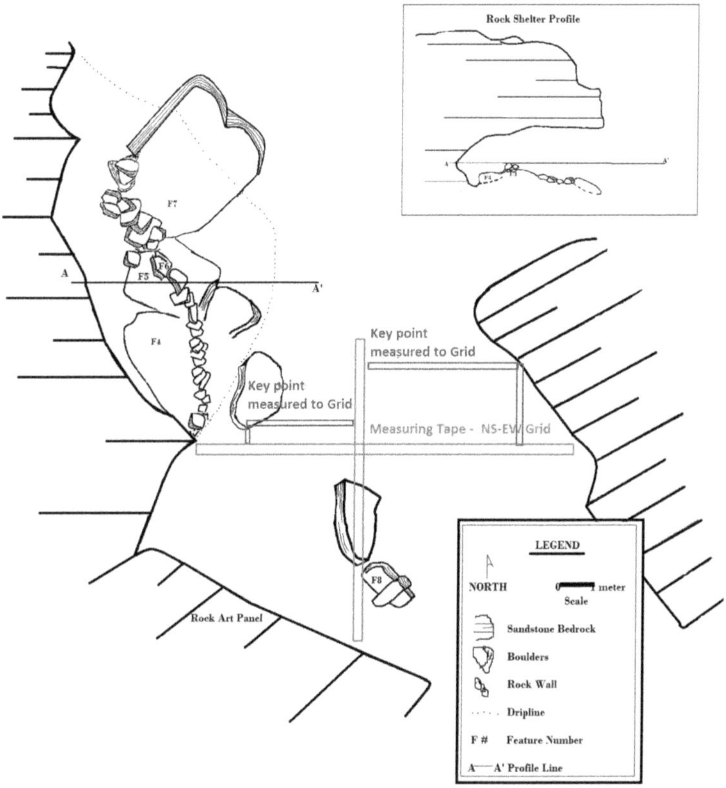

Figure 4.1. Planview map of 5LA11446 showing main features and their spatial relationship. Created by Elizabeth Lynch.

Figure 4.2. Overview of site 5LA11446. Rock Art panel is blocked out. Photo Credit: Lawrence Loendorf, used by permission.

Figure 5.1. Overview of Metate Cave from the creek. May 2011. Photo Credit: Elizabeth Lynch.

Figure 5.2. View of landscape from inside Metate Cave, October 2016. Photo Credit: Tom Noble, used by permission.

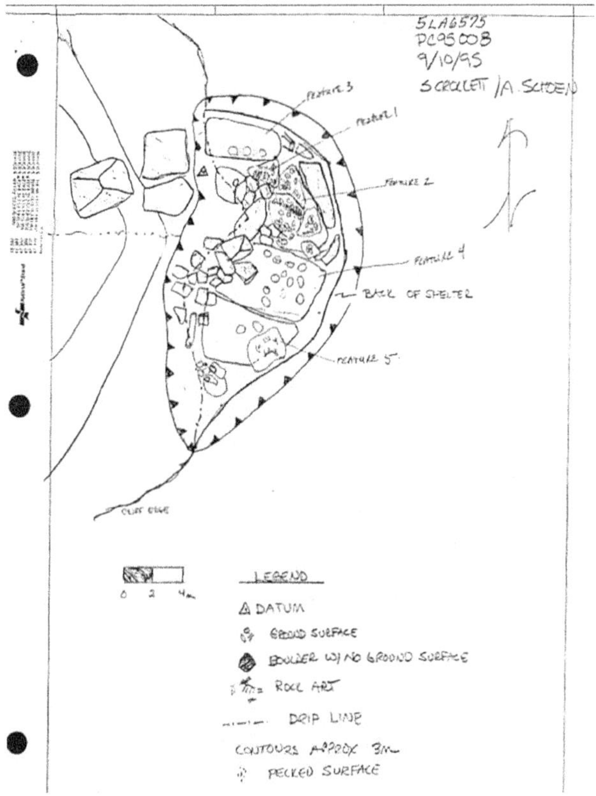

Figure 5.3. Field drawing from 1995 Survey Photo Credit: USAG Fort Carson, used by permission.

Figure 5.4. Model of all the features within the Metate Cave (5LA6575) rockshelter. Photo Credit: Elizabeth Lynch.

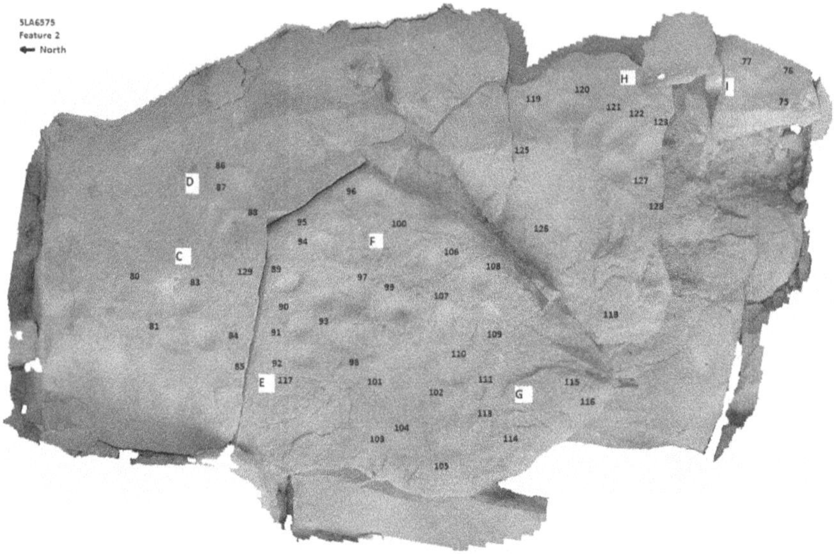

Figure 5.5. Feature 2 at 5LA6575 with elements (numbers) and sub-features (letters). Note the placement of sub-feature I. Photo Credit: Elizabeth Lynch.

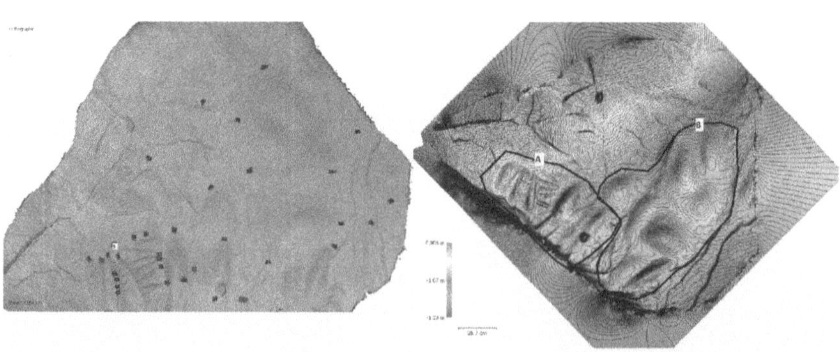

Figure 5.6. Feature 1 illustrating the element identification from the 3D model (left) using field tape markers; sub-feature designation from digital elevation model (right) illustrates the two possible sub-features discussed in the text. Photo Credit: Elizabeth Lynch.

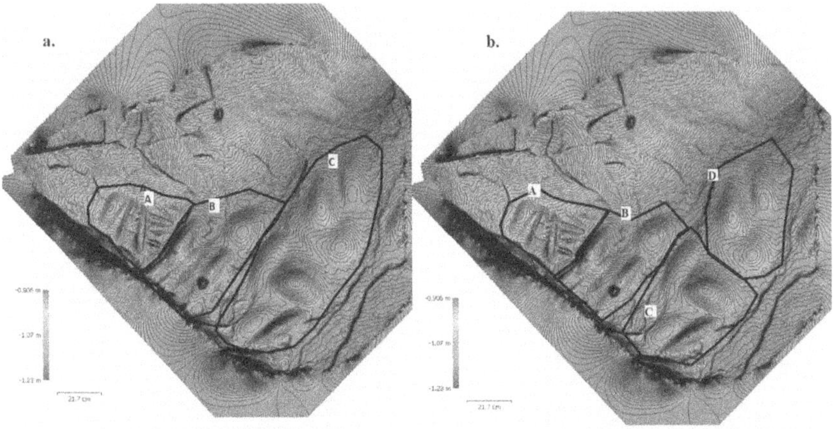

Figure 5.7. Illustration of possible scenarios for sub-feature designation using the digital elevation models to visualize the use-wear: a) three sub-features with a mixture of element types; b) four sub-features using proximity and similarly shaped elements. Photo Credit: Elizabeth Lynch.

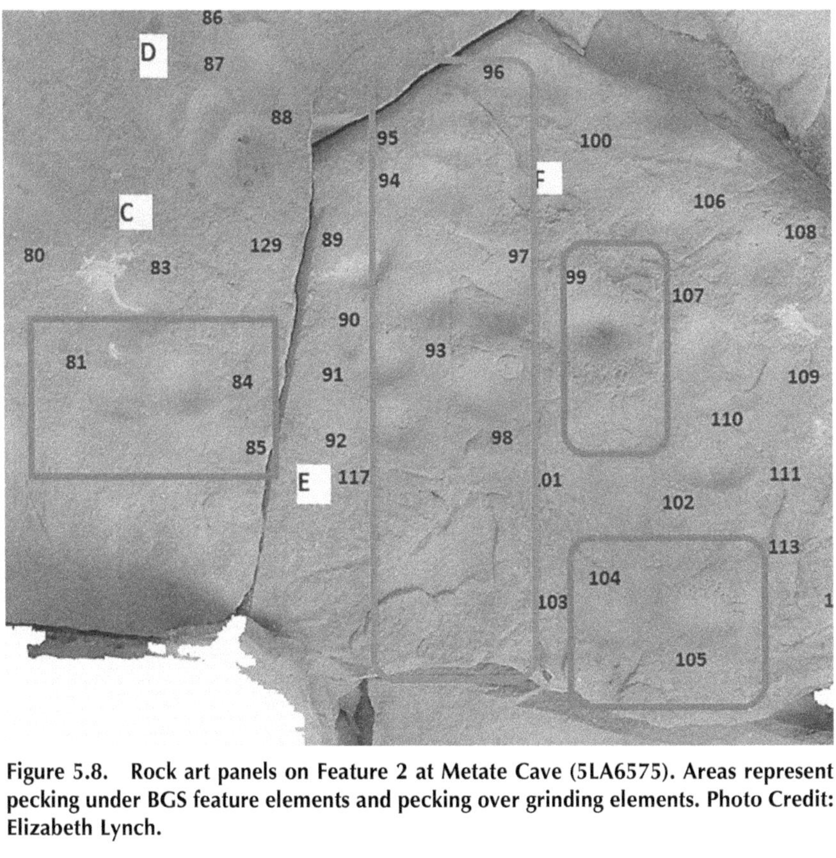

Figure 5.8. Rock art panels on Feature 2 at Metate Cave (5LA6575). Areas represent pecking under BGS feature elements and pecking over grinding elements. Photo Credit: Elizabeth Lynch.

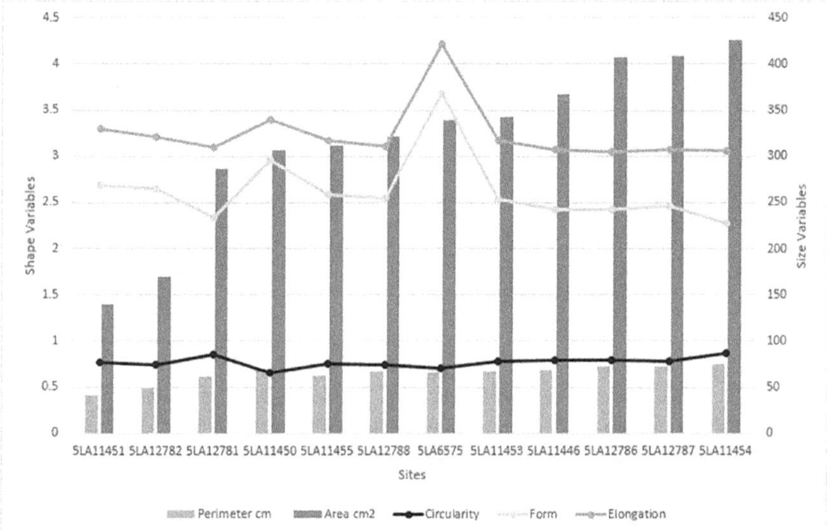

Figure 5.9. Illustration of the difference between all shape characteristics between sites located in the JE Canyon Survey and 5LA6575 (the peak). Shape variables below zero are circularity and elongation (1 is a perfect circle) above 1, and the shape variable is form. Photo Credit: Elizabeth Lynch.

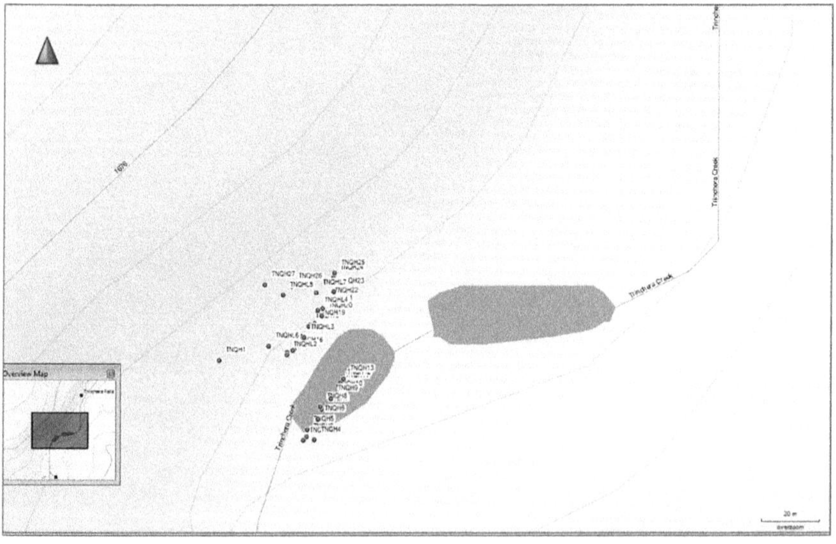

Figure 6.1. Map of the 360-degree camera and LiDAR locations used to create the Virtual Reality Tour. Photo Credit: Elizabeth Lynch.

Figure 6.2. First stitched 360-degree image of Trinchera site at the creek, rockshelter can be seen on the left of the image. Photo Credit: Elizabeth Lynch.

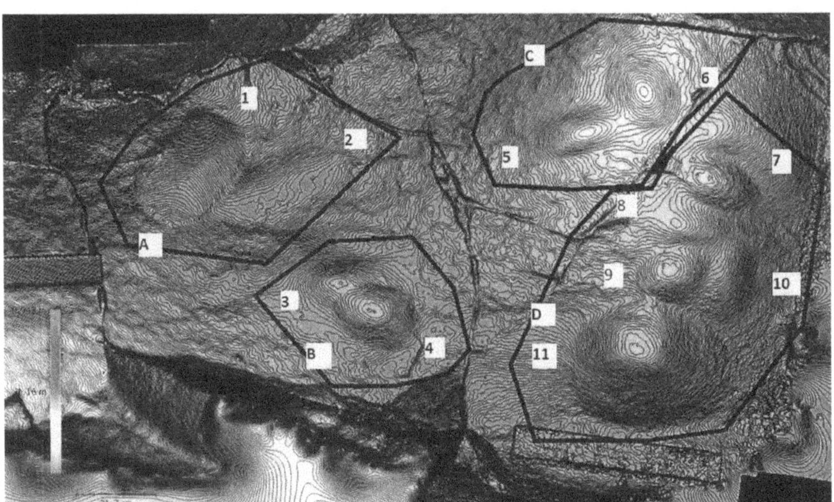

Figure 6.3. Digital elevation model with contour lines of the main BGS feature at Trinchera Cave. Sub-features (letters) and elements (numbers) are labeled to illustrate the different grinding element orientations. Those within the subfeature boundaries are related by proximity. Photo Credit: Elizabeth Lynch.

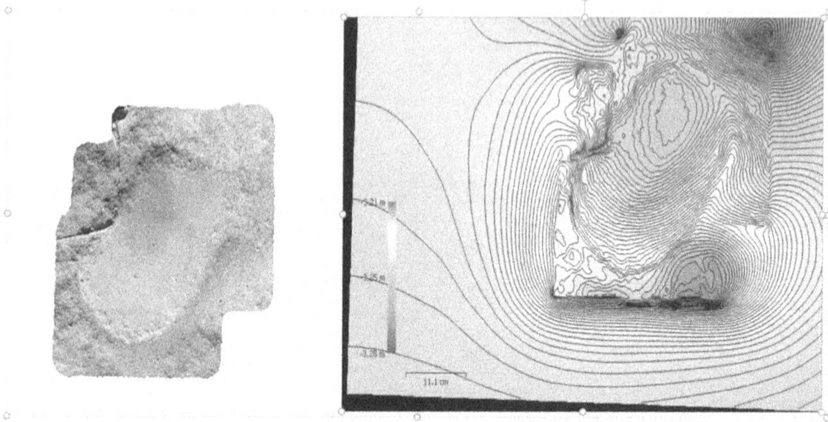

Figure 6.4. Sub-feature A, element 1, a trough-shaped grinding element similar to those found in the Southwest. The model of the grinding surface (left) and the digital elevation model (DEM) with contours is on the right. Photo Credit: Elizabeth Lynch.

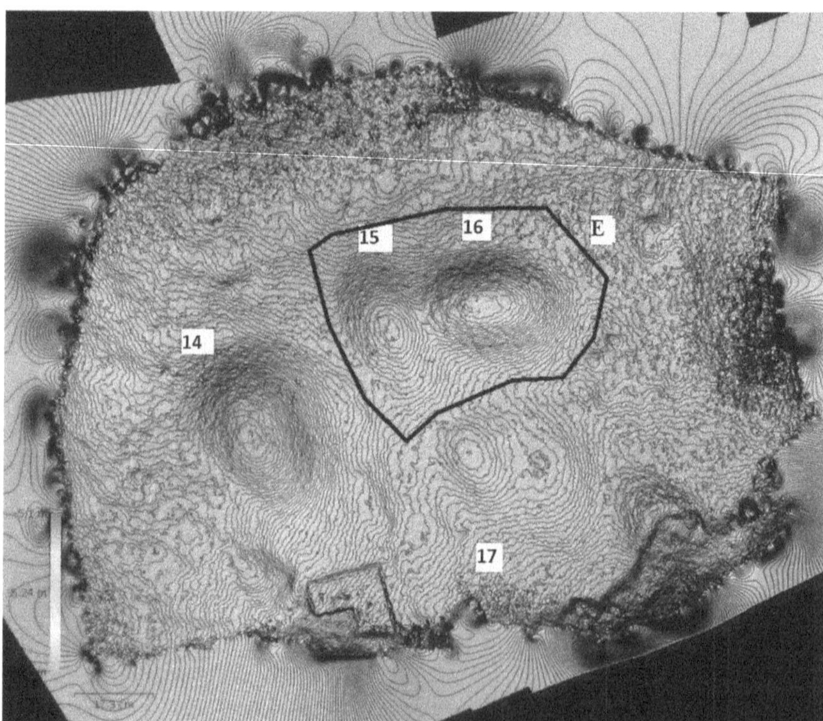

Figure 6.5. DEM with Contour lines of Feature 4 showing sub-feature A and element numbers. Photo Credit: Elizabeth Lynch.

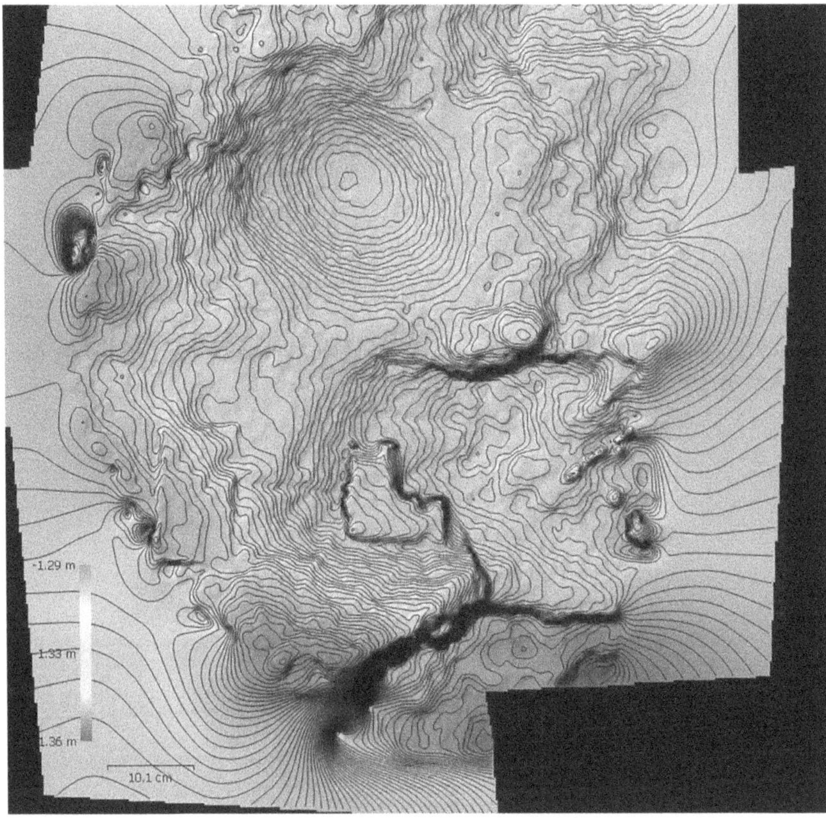

Figure 6.6. DEM with contour line of Feature 3. Photo Credit: Elizabeth Lynch.

Figure 7.1. Example of an open-air Paleoindian site. A & B. Photo Credit: USAG Fort Carson, used by permission.

Figure 7.2. Example of an open-air Archaic Site A& B Photo Credit: USAG Fort Carson, used by permission.

Figure 7.3. Example of a Prehistoric open-air feature. Photo Credit: USAG Fort Carson, used by permission.

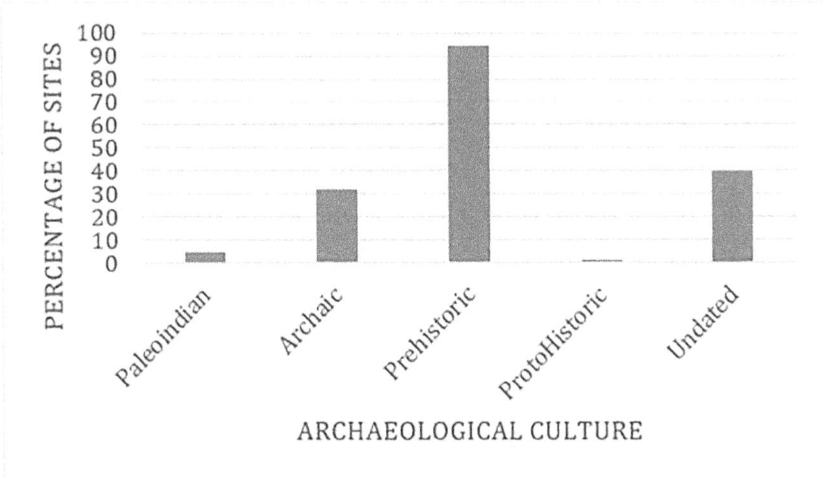

Figure 7.4. Frequency of BGS feature sites through time on the Chaquaqua Plateau. Created by Elizabeth Lynch.

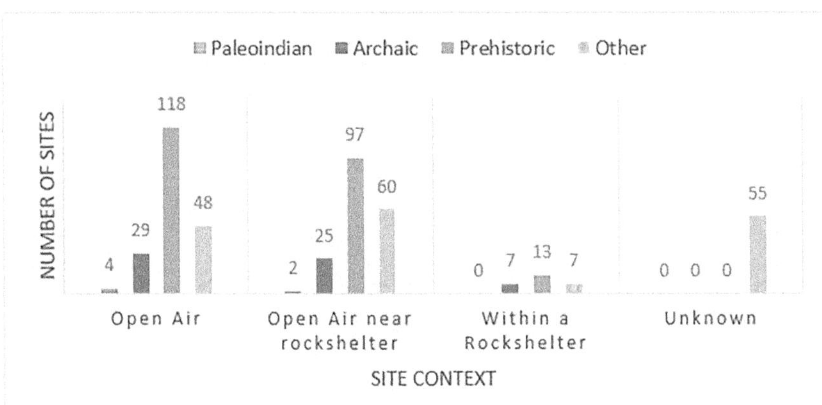

Figure 7.5. Frequency of Landscape Contexts for BGS feature through time on the Chaquaqua Plateau. Created by Elizabeth Lynch.

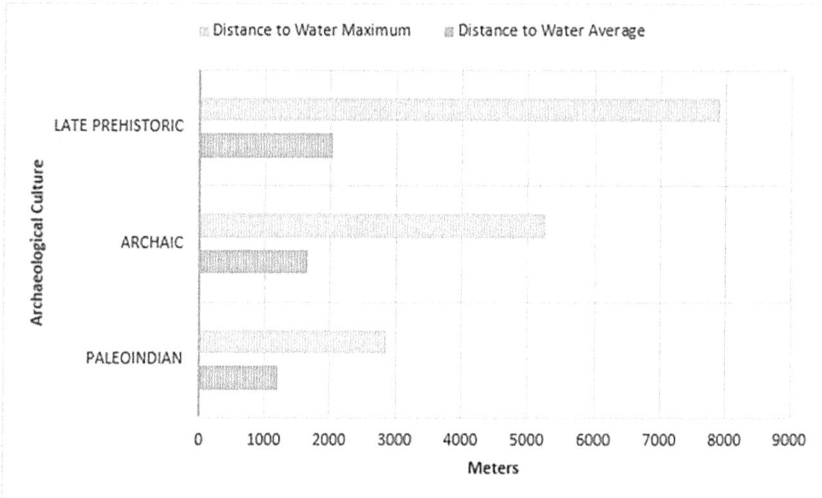

Figure 7.6. Distance of BGS feature sites to water, the average, minimum, and maximum distance are in meters. Created by Elizabeth Lynch.

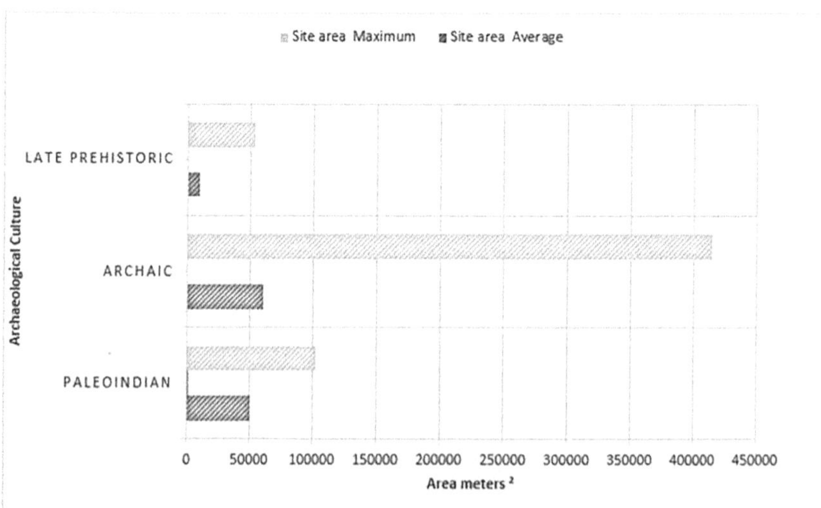

Figure 7.7. The frequency site size (area in meters²) of BGS feature sites through time on the Chaquaqua Plateau. Created by Elizabeth Lynch.

Figure 8.1. *The Mealing Trough*. 1906. Edward S. Curtis. Four Maidens in Native Dress, Grinding Corn in Milling Bins with Manos and Metates Inside Adobe House; Meal Brush and Baskets Nearby. Photo Lot 59, LOC, Small Mounts, Tribe Id, Pueblo, Hopi, Food 03317900, National Anthropological Archives, Smithsonian Institution. NAA INV 08547317, NAA MS 4731.

Figure 8.2. *Women Grinding Corn and Baking Bread*. 1898. Mooney, James. Pueblo of Hano on Hopi Mesa, Arizona. Negative 1878 A 1, National Anthropological Archives, Smithsonian Institution.

Figure 8.3. *Woman Pounding Acorn with Children Nearby.* National Anthropological Archives. Smithsonian Archives.

Figure 8.4. *Woodcutting of Women Pounding Acorns* by Joline J. Butler (1815–1846). National Museum of American History, Kenneth E. Behring Center ID Number: 1999.0145.207 accession number: 1999.0145 catalog number: 1999.0145.207. Permission pending.

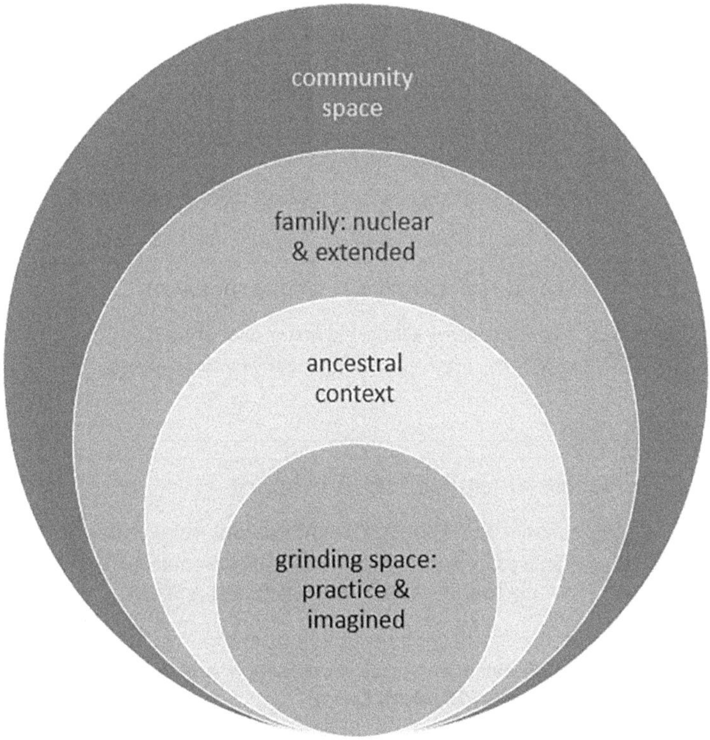

Figure 8.5. Figurative drawing of grinding space with key actors and material culture. The space is active and participatory with past and present; the space is both imagined (thought about and talked about even when grinding is not taking place) and part of material reality. Created by Elizabeth Lynch.

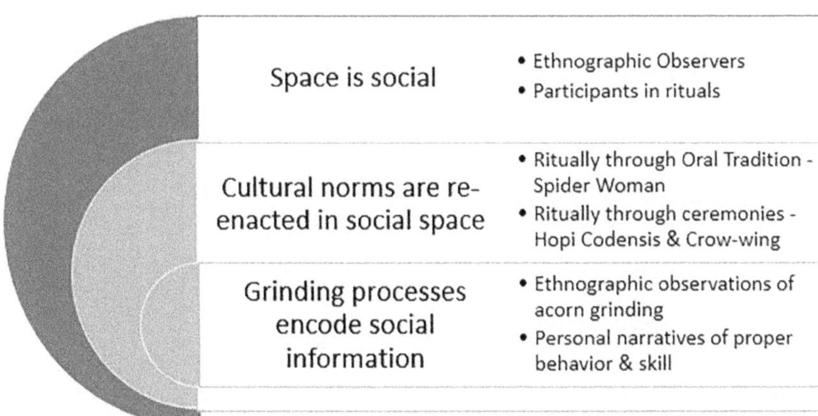

Figure 8.6. Multi-vocal perspective of grinding workspaces. Created by Elizabeth Lynch.

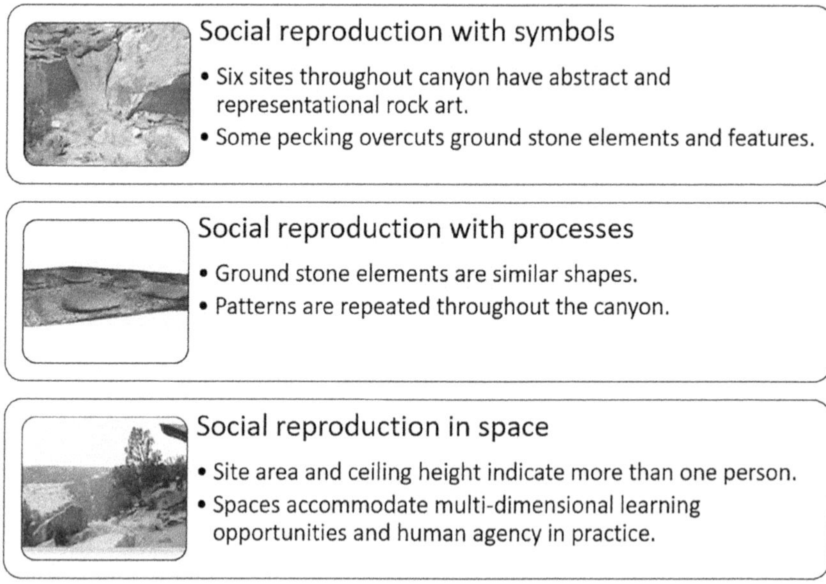

Figure 8.7. Summary of the main social reproduction processes observable in the Chaquaqua Plateau. Created by Elizabeth Lynch.

6

3D Visualization of BGS Features

Trinchera Cave Case

Recent advancements in digital field recording methodologies include the use of photography to create visual representations of the site in multiple dimensions and virtual reality. These virtual models are very useful for engaging with sites that are difficult to access or with those, like BGS features, rock art, or architectural sites where the features remain *in situ*. Perhaps one of the more sophisticated applications of this type of technology is the ability to create virtual collaboration between researchers, descendant communities, and archaeologists, as well as creating outreach and educational options for the public.

In this chapter, we apply recent developments in the digital collection of archaeological data using photogrammetry and virtual reality methods for recording the BGS features at Trinchera Cave (5LA1057). Trinchera Cave is an archaeological site critical to understanding interactions between regional cultural groups, as evidenced by the presence of material culture from the Southwest and Southern Plains (Wood 1974). The site is very accessible and well-known, so not surprisingly, it is considered a disturbed context. Bedrock ground stone at the site has not escaped the effects of such attention, so digital recording methods offer the opportunity to explore how to record displaced BGS features while also providing means to preserve the site context through virtual reality.

One of the questions often posed, when applying digital methods is discussed, is: why do we need these new methodologies? The manual recording techniques are well-known and field-tested. The equipment needed is simple, light, and versatile. The lab protocols for processing the manual records are well-established. In short, we can get all the data we need to describe, analyze, and compare regional bedrock ground stone features and elements using currently established procedures.

That is all true. Previous chapters have illustrated that while certainly old-school, these methods are still indispensable for thoroughly documenting BGS features. However, if we hope to engage a broader audience in BGS research, we need to replicate findings between research projects. Dialogue needs to be generated and nurtured amid diverse perspectives about past material culture. We can do this by bringing together the often difficult-to-reach sites with the publics' imagination while also protecting traditional cultural resources. Manual recording techniques have high interobserver bias. The only way to ameliorate our databases is through engagement by multiple researchers with often disparate points of view. I show that creating digital models (data sets) accessible to interested persons means that we can both minimize bias and increase our ability to replicate each other's work. At the same time, we minimize traffic to the sites themselves, thus preserving them for future generations. Archaeologists can produce applications or 3D printed models for museum exhibits or develop completely digital site tours, thus engaging with broader communities of learners.

SITE BACKGROUND

Trinchera Cave was documented and excavated by Haldon Chase. Excavations have continued through the years (see Zier 2017 for a definitive history of the excavations at the site). The site has not been well-documented and is now part of a district of sites included on the National Register of Historic Places. Missing from what scholarly work has been done on the material culture from the Trinchera site, is a description of the site's bedrock ground stone features. These features are prominent components of the site. Indeed, early archaeologists worked around them, possibly further disturbing their locations (see illustrations in Zier 2017). Early visitors even photographed the features (Figure 1.3). Yet, they remain excluded from the mainstream discussion about the site, except as evidence that food-grinding took place here. The general inconspicuousness of these features may be the side-effect of salvage archaeology. The role of salvage archaeology projects is to collect and save as much material culture as possible before they are permanently lost. Digital recording methods can position BGS features within other discussions about the site inhabitants.

Because of the site's location, there are occupations from Plains cultures and influences from the Puebloan world to the south and southwest. Trinchera is one of the few sites that we know of, on the Chaquaqua Plateau, with the trough-style metate ground into the bedrock. Trough-type BGS feature elements have been documented in the four-corners region (Villalobos-Boehm 1997) but are rare in southeastern Colorado. The BGS features are in close association with sizeable intricate rock art panels. There are multiple types of grinding elements, but tool grooves are missing. The site offers opportunities to add to our knowledge of basin-style ground stone changes through time, particularly as the region became influenced by the Puebloan cultures to the west. Trinchera Cave may be a cultural melting-pot of sorts. Does cultural diversity of BGS features reflect tolerance of "other," or are they from different time periods, therefore not reflecting dynamic cultural contact? If there was a continual cultural exchange in this area, how did it affect other inhabitants of the Chaquaqua Plateau, particularly those at JE Canyon or Metate Cave?

CHRONOLOGY

The excavations conducted within the Trinchera rockshelter nearest the BGS features by Wood-Simpson in 1974 yielded a debated chronology for the site. She discussed a date range of 775 BP to 465 BP, based on pottery and radiocarbon dates. A Meron-type projectile point suggested to Wood-Simpson (1976) a long-term occupation in the shelter dating to the Archaic. The Late Archaic falls between 3000 and 1850 years before the present (BP). Additional excavations by Colorado College (Nowak and Gerhart, 2001 and 2002) provide a range of radiocarbon dates between 1100 to 750 BP with an outlier of 2760 to 2640 BP. Bartlett's (1933) and Eddy's (1965) timeframe for the shift between oval basin-like ground stone to trough-like metates is at the end of the Archaic about 2,000 years ago. Bartlett's analysis illustrates that the trough style becomes dominant at sites in the Southwest about 900 years ago.

ENVIRONMENT

Trinchera Cave, really more of a rock overhang, was created by the erosion of the Dakota Sandstone bedrock shelf. Roof fall boulders are scattered on the floor of the shelter. It is on these boulders that people created ground stone surfaces. The shelters have thick deposits of sediments that have accumulated through time (Zier 2019). Trinchera Creek runs through the site within walking distance. The environment is similar to the landscape at Metate Cave. However, at Trinchera, the rockshelter is more open. The site is accessible from the creek, but there is a natural embankment running the length of the shelter that creates a somewhat sheltered feel. Nonetheless, some of the erosion to the BGS features may have been due to wind, water, and freeze-thaw processes.

The bedrock ground stone features have been badly damaged from natural erosion at the site and through constant human traffic in and around the boulders. Additionally, some of the original boulder surface areas have broken, leaving disassociated grinding elements near but separate from the main BGS feature. Another substantial boulder with ground surfaces lies nearby. It is obviously out of place, either from settling of the softer sediments or perhaps moved there by earlier excavations (see Zier 2017 site map for the placement of site BGS features). Although it is probably not *in situ*, some elements are visible. Close-range photogrammetry (CRP) can be used to develop 3D models of the features for analysis and inclusion in a regional database of BSG feature elements. The local plants resemble those found elsewhere in the canyon systems. The presence of rich alluvium makes horticulture feasible near the rockshelter (Zier 2019).

METHODS

The BGS features at the site had not been fully documented at the time of our visit. However, a graduate student at the University of Colorado Denver was in the process of studying the ground stone assemblage from the site, including the BGS features (Ryan Baker 2018, personal communication). The goal of our visit was to capture the landscape surrounding the rockshelters for the development of a virtual reality tour of the site and capture the rock art and BGS features using photogrammetry. The team consisted of four undergraduate students from the University of Wyoming Digital Archaeology course and a visiting graduate student from Utah State University. We had received training in LiDAR and 360-degree camera equipment from our partners at the UW Shell 3D Visualization Center, University of Wyoming. Since we had only one day to capture as much data as possible, we split into two teams: one to record the landscape the other to record the features in the main rockshelter. All models were processed at the Shell 3D Visualization Center, by members of the project team and by Kyle Sommerfeld (3D Viz Center).

RECORDING METHODS: CLOSE-RANGE PHOTOGRAMMETRY

The rockshelter proved to be a relatively large project for training purposes. Two students worked together to capture overlapping photography of the main rock art panels, while the author captured data sets from the BGS features. We used two Canon EOS DSLR cameras. The cost of calibrated target boards might prove a deterrent for nascent photogrammetry projects, so we experimented with means to capture data using homemade target boards from images found in open contexts on the internet. The homemade calibration methods might produce model error beyond some project requirements, so the pre-calibrated target boards are highly recommended.

In keeping with the goals of standard photogrammetric practices, we placed target boards on or near the subject of interest but not covering the elements. Since the field trip was part of student training and since the rock art panels were reasonably large projects, most of the field time was spent practicing the proper layout of the camera shots and reviewing for 66 percent overlap of images. We assigned temporary feature numbers in the field. Grinding element identification and subfeatures were assigned during lab processing.

The prominent bedrock boulder with BGS features in the rockshelter was divided into two different features based on the topography of the boulder. Feature 1 has the only trough-like surface on the site. Feature 2, located on the same boulder, is near a pecked rock art panel. Feature 3 is a standalone BGS feature that may have once been part of the prominent BGS feature boulder but is now partially buried along the shelter's floor. Feature 4 is a single grinding element located on a rock, cleaved from the main feature at some point in the past. The latter was recorded as an afterthought but provided some surprising information in the lab.

RECORDING METHODS FOR VIRTUAL LANDSCAPE TOUR

A single student-led team was assigned to capture both 360-degree shots and LiDAR shots to construct a virtual landscape tour. The entire class spent about an hour touring the area, then decided on the best layout for cameras. The project started in the flood plain area outside the rockshelter, then traveled downstream and back upstream to collect 360-degree photos and LiDAR (Figure 6.1). The 360-degree camera was placed in areas that afforded a continual line of sight between camera shots. To capture as much of the immediate environment of the rockshelter, we also took 360-degree image captures along the top of the ridge above and adjacent to the site. The 360-degree camera takes six photographs simultaneously. These are then stitched together in the lab (Figure 6.2). A virtual tour was created in Unity from the 123 stitched images captured in the field. The VR tour application is housed at the University of Wyoming 3D Shell Visualization Center. Unfortunately, the LiDAR project did not yield helpful results.

Building 3D models

Models of the BGS features were created with Agisoft (now Metashape) modeling software. The general workflow creates the model, reduces error, introduces a scale, produces a digital elevation model (DEM) and contours, captures image data, and exports a shape model (obj or ply files). The data will ultimately be part of a digital asset library so that other researchers can download them. Thus, the digital photographs used to build the models were stored in project files using processes adapted from Noble and Matthews (2018) and Cultural Heritage Imaging workflows. In this teaching and learning field project, we experimented with using raw and jpeg settings (as in the previous case) and using only jpeg data. The latter is a salvage technique. The image data captured has data visualization applications, but the models have minimal practical use at the submillimeter level of measurement. The results of the jpeg models are reproduced below.

Using just jpeg image capture can be a good tool for quick surveys, where checking field camera settings might be overlooked. The barest minimum units needed are good quality photographs with 66 percent overlap and a scale. While smartphones can be used to capture the photographs, there will be a large amount of error in the resulting models if the autofocus is on. However, these will still produce good visualization products and models that can reproduce critical aspects of the BGS features in the site context. The error can be overlooked to capture a 3D model of the surface area for descriptive purposes. The entire documentation process from field capture to archives can be found in Noble and Matthews (2018).

RESULTS

The UW field trip documented four BGS features with five sub-features and seventeen elements. We concentrated our efforts on those features that were the most visible, therefore under more stress from visitors to the site. Measurements and morphometric analyses of the elements were entirely produced from the lab models (publication forthcoming). Here we illustrate how BGS features can be minimally described and reproduced using only photogrammetric methods.

Feature 1 is the most visible BGS feature on the site located within the rockshelter (Figure 6.3). It is a very protected area that overlooks Trinchera Creek and the rest of the rockshelter. The feature is surrounded by the rock art panels covering the walls of the shelter. Motifs range from older pecked motifs to protohistoric motifs like shields and horse-like images.

Feature 1 has eleven associated grinding elements. The 3D model, generated in Metashape, used jpeg images to produce a digital elevation model (DEM) with contour lines. The feature surface has multiple slopes. The pattern of the elements into groupings of seeming paired elements is immediately visible. However, this twinning pattern is not unique to Trinchera. The pattern also appears in the JE Canyon survey as well as at Metate Cave. The two main hypotheses of such patterns are the human-defined work areas that include multiple processing stages or that the pattern represents a form of use/abandonment where one element is abandoned (for whatever reasons), then a second area near the first is chosen to begin processing again; thus, it represents a form of sequential behavior. Designing tests of these scenarios is in the nascent stages.

The trough shape, Feature 1, element 1, is unique among the other elements. It is at an angle that crosses over and truncates the edge of element 2, thus most likely post-dates element 2 (Figure 6.4).[1] Further, the contour lines can help us determine the position of the grinder. If the person who used the space needed an angle to create grinding pressure, then we expect the deeper, steepest part of the element would be distal (further away from) to the grinder's body. The pattern is represented by the darker shading where contours are more closely spaced together, indicating a steep slope. The wider spaced contours represent a more gradual slope. In this scenario, the proximal (closer to the grinder's body) ends seem to start in other grinding elements, which further supports the hypothesis that grinding element 1 is superimposed over grinding element 2.

The sub-features A and B appear to represent two separate periods of grinding with elements that crosscut one another. Sub-feature A elements 1 and 2, discussed above, exhibit this pattern. Sub-features B, element 4 appears to have occurred later than element 3. Sub-feature C, elements 5 and 6 are oriented perpendicularly, so may have been used together or at different times. Sub-feature D elements 7 through 11 are so densely packed that it is difficult to break out their relationships. However, unlike the other elements on this feature, elements 7, 8, and elements 8, 10 are oriented on parallel axes, which seem to be oriented in the same direction, as evidenced by the contour lines.

The interpretation of Feature 1 elements suggests that one or two persons worked in the space at any given time. While multiple-step processing could have occurred at sub-features C and D, it is also plausible that single workers created the wear pattern we observe today. One other important aspect of this feature is surface pecking. Often interpreted as retouching of the ground stone surface, creating a roughened surface for grinding might also reflect weathering processes. However, results from the Metate Cave 3D models suggest that there may be some grinding overlaying pecked rock art motifs.[2]

Feature 2, located on the same parent boulder as Feature 1, is separated by the shape of the boulder. It is a single elongated element (Feature 2, element 12) that neighbors a heavily patinated rock art panel (Figure 6.5). The basin is evenly ground but still shows a distinct proximal end that angles towards a deeper end, although it is not as prominent as the neighboring elements of Feature 1. The grinding element is on a parallel axis to the rock art panel, but a naturally occurring ridge separates the two. The rock art panel is a series of intersecting lines with small cupules or small (1–2 cm diameter) pecked or ground circular elements. The cupules could be pecked and related to the rock art or caused by shotgun or other firearms. Only further study of this panel will help to place the grinding element into site context.

Feature 3 is a small standalone boulder at the base of Feature 1 (Figure 6.6). Element 13 is a relatively deep, evenly ground surface with ancient rock art motifs along the basin's edge. Likely, this boulder was once part of the prominent boulder when both the basin and the rock art were created and used. The cracks in the parent boulder predict where additional erosion will occur along natural fracture lines. Feature 4 is a relatively sizeable stand-alone boulder partially buried in the heavily excavated and trafficked portion of Trinchera Cave.

Historic photographs indicate that it is possible the excavators themselves may have caused the movement of the boulder. However, because of its size, it may have been shifted due to natural processes. We were not authorized to excavate or move artifacts around, so there are probably more elements on this BGS feature. The feature has at least four elements (14–17) that are grouped into three workspaces, one sub-feature (E) with the same pair-grouping found on Feature 1, sub-feature C. Because this feature is not *in situ*, individual basin models were developed to determine the proximal and distal ends of the ground stone elements. Element 14 is a relatively deep oval-shaped basin. When the model is oriented perpendicularly, there is little evidence to determine the proximal end of the basin. However, there is a slight bulge on the lateral edge, which may evidence the movement of resources between the sub-feature A and this basin. They are close neighbors, and the bulge points towards the sub-feature. The "bulge" is also seen on the sub-feature DEM. In this instance, the proximal end of grinding element 15 is visible, while the proximal end of 16 would originate in element 15, thus suggesting that 15 predates 16. It is plausible that 14 and 15 were used in concert by different grinders, then as time progressed at the site, basin 16 was created to be used alone or as part of a process involving basin 15. Basin 17 separate from the other areas, but you can see the relationship to element 15. This element is shallower than the others, but the proximal end is just barely discernible.

BUILDING A VIRTUAL REALITY TOUR FOR SCIENTIFIC VISUALIZATION

The process of building 360-degree tours of a site has been simplified since this recording as technology continues to make the smartphone a do-all device. For instance, there are 360-degree camera attachments with apps that can construct virtual reality tours on the fly. These tools are handy and easy to learn. The goal of our fieldwork was primarily exploration and training a new generation of archaeologists in the process of landscape modeling and development of the archivable virtual reality environments to preserve and describe archaeological sites.

SCIENTIFIC VISUALIZATION AND 3D MODELING

The use of close-range photogrammetry (CRP) and 360-degree methods to record sites provides a scientific means to better understand patterns in the archaeological record while also preserving the integrity of sites for future generations. The process is no more time-consuming than standard recording procedures, with properly trained individuals, and provides a rigorous means to test multiple hypotheses about BGS features. During the analysis phase of this case study, the virtual site was used to re-visit the fieldwork and test the resulting lab interpretations to fact-check BGS feature locations.

Although CRP can create very rigorous data sets for comparative studies, at the fundamental level, they provide the means to describe site features accurately. While manually recorded field measurements can be precise, there is also potential for bias or observation error. Photogrammetric models, on the other hand, provide a quick means to record a feature faithfully. One problem that needs to be further studied is the standard protocol for creating the models and the best means to interpret the slope of the features and elements. While sub-features seem to be a straightforward way to describe group work, elements on Features 1 and 4 suggest that the interpretation process needs to be more rigorously tested.

Data Management

The use of 3D modeling as a method for archaeological analysis leads to a challenge in data management. What exactly happens to the models? What should be archived? Who controls the virtual tours? How is this data to be managed in long-term archives? What should be made public, and who gets to decide access? In many ways, these are issues of storage and sustainable data management and fall in the fundamental ethics of what we do as archaeologists and custodians of the past. A tangential issue concerns the creation of narratives about the past in cyberspace. Are we creating a new scenario of colonization in which Native American communities are further excluded from the production of knowledge about the past intricately related to their Traditional Cultural Properties?

In this case study, as in most of the work completed in this book, the data are managed at the researcher level. At this point, the models and underlying data sets are housed within the researcher's archives on personal external hard drives or university servers. The data are available only to the site stewards, Native American descendant communities, or approved researchers. The data management plan for the archives includes creating a sustainable data set with models (and a workflow) that is accessible through standard Microsoft 3D software in the absence of access to Metashape or other 3D modeling software. The agency or steward of the property is responsible for the long-term data management of their archive. At the same time, the researcher maintains control over the original dataset, models, and workflows at her home institution.

Publication of data undergoes a review process by the institution but is rarely subject to consultation. The question is whether virtual reality tours should be made public without a consultation process. This ethical dimension is only just being considered by the scientific and Native American communities; therefore, it is still evolving.

CONCLUSION

We have progressed to the stage of site recording that analyzes bedrock ground stone features more user-friendly than ever. Implementation of the CRP, 360-degree, and LiDAR site recording is very affordable. Modeling software, although still relatively expensive, is widely available, user-friendly, and perhaps an indispensable part of modern field recording. While producing virtual reality tours for public consumption falls into an undiscovered country of ethical concerns which need to be discussed and confronted in the forum, we find researchers engaged with the fieldwork but not releasing the applications until full consideration has been given to the consequences of this type of publication. That said, the models foster a new realm of scientific visualization in which the research interest in BGS features is no longer dependent on two-dimensional notes but can now generate 3D models to both visualize and describe past behavior. The analysis and interpretation of which will require additional experimentation and consultation.

NOTES

1. The superimposition coupled with dates from the site indicate that oval basins were used before the trough shaped ground stone. It seems reasonable to interpret the latter shape as an intrusive cultural element. It also seems likely that the Plains grinding element pattern of oval-shaped ground stone may have been in use concurrent with and after the trough-style grinding element.

2. The suggestion that pecking on a surface indicates retouching of the surface, is open to debate. Although ethnographic examples of this practice exist, many times the grinding element is under or near a dripline. While making this assertion for portable ground stone seems logical, those found in rockshelter or exposed to the open sky, might represent weathering. Dakota Sandstone is very friable, so we should be cautious in making such blankets statements without additional evidence to support the supposition. The case is another good reason for drawing scaled site maps that indicate potential for natural erosional processes.

III

RECONSTRUCTING LANDSCAPE USE AND SOCIAL SPACE

7

Inclusive Landscapes through Time in Southeastern Colorado

Archaeology is powerful because of the capability to operate with diverse material cultures from the deep past. Archaeologists are well-equipped with methods to place artifacts into time sequences. We use both relative and absolute dating methods to accomplish this task. Unfortunately, BGS features are resistant to absolute methods such as C-14 dating, and while some cosmogenic methods might work, the sampling processes are highly destructive. Since many of the BGS features sites might be related to past ritual or socio-sacred events, non-destructive, relative methods using superposition and seriation are more valuable tools for placing these features tentatively within the broader regional, chronological context.

Relative dating requires several assumptions. The most important is that the BGS features are related to other datable materials that are nearby. These materials can be hearths features (which can be dated with absolute methods like Carbon-14), rock art panels (which can be dated using seriation methods), architecture, and diagnostic artifacts (such as projectile points and pottery). Placing the use of BGS features into the local chronology requires the use of all these methods as well as multiple lines of evidence. The results are circumstantial at best. We cannot know if the grinding features were used simultaneously with the rock art panels or in conjunction with a hearth feature. Nonetheless, material culture found together at the same location *is* spatially related, so that is our starting place.

Excavations at Trinchera Cave provided many reliable dates for the cultural activities through time. Since there are BGS features within the rockshelter, they are probably associated with at least some, if not all, of the events in its life history. Absolute dates indicate the site was occupied between at least 1110 and 750 years before the present (BP), with an outlier date to the Archaic Period (about 2700 BP) (Zier 2019). Relative dates (using the artifacts from excavations) range between circa 850 to 500 BP. Therefore, the BGS features might have been created any time between 2700 BP and the most recent date, 500 BP. The outlier date (2700 BP) is used to remain conservative. The earliest date is probably closer to 1100 and 850 BP. If the rock art panels at Trinchera were analyzed, like those of JE Canyon Ranch, we could use the same logic. The rock art panels at Trinchera Cave are likely to have been added to well into the historic period. However, some are very old pecked motifs found near and under the grinding elements with over-pecked designs like those at Metate Cave. So how do we know which dates related to the BGS features?

At Trinchera, there is evidence of at least two different material cultural influences (Plains and Southwest). The BGS feature elements are typical of Chaquaqua oval BGS grinding elements and Southwest trough grinding elements. If Bartlett (1933) and Eddy (1964) are correct, the trough-metate shape begins about 2,000 years ago in the Southwest but is a dominant design by about 900 years ago. The trough element slightly overcuts an oval grinding element. Thus, the trough element is more recent than that element. So, it is not unreasonable to place the BGS trough-element between 2,000 and 900 years ago. The BGS features at Trinchera were probably first used sometime during the Late Archaic and persisted until about 500 years ago, since the trough shape persists to modern use.

Another assumption is that the features are somehow related to similar features throughout the region, which means that the people are somehow related in terms of socio-cultural norms. Similarities in rock art motifs, artifacts, and architecture support the idea of related archaeological material culture, supporting the notion that the people making and using these items are also related. Thus, we have to assume that the people who used the oval-shaped elements at Trinchera (at some point in time) were somehow related to those who used features with similarly shaped elements at JE Canyon and Metate Cave. They may not have been closely related, but at least followed similar cultural traditions. However, the trough-like element at Trinchera has not been found at these other sites. Therefore, at some point in time, the cultural similarities changed. Of course, this sequence of events and assumptions must be constantly questioned and tested. The oval-shaped basin is an early design in the Southwest (Bartlett 1933; Eddy 1965) so, relying exclusively on general shape categories without a corresponding morphometric study remains speculative. The BGS features at Trinchera provide a good opportunity for morphometric research into the evolutionary pattern of BGS elements.

The example presented should illustrate why it is essential to attempt to place BGS features into a time sequence, no matter how broad. Understanding how the features relate to each other and other archaeological, cultural periods helps us understand how they were used and why they were abandoned. Were the BGS features more important during relatively stable climate periods or when environmental stress occurred? Is there a gradual evolution in how the features were used, or did this type of behavior remain relatively stable throughout time? If the same group of people (related by the landscape, cultural behaviors, and perhaps language) used the features, and if they inhabited this region for thousands of years, they are very well-adapted to the regional environmental changes. So why would the grinding technology change? What advantages made this technology and the socio-cultural behaviors associated with it relatively stable? With an idea of the types of variation in BGS feature sites, we can begin to look at the variation in material culture most commonly found at BGS feature sites.

In this chapter, we are concerned primarily with the preliminary attempt to place BGS features into a chronological sequence that reflects the primary archaeological sequence for the region as outlined by Zier and colleagues (1999). While there are undoubtedly other regional, chronological sequences, this one (Table 7.1) is used by most cultural resource managers as well as the Colorado Rock Art Association (CRAA). It is the primary time context used by regional archaeologists. Thus, the most straightforward path is reconciling the features with an existing, well-developed framework.

The GIS database of all sites with bedrock features, provided by the OAHP, had 459 sites in the region with bedrock ground stone features (see chapter 3). Regional archaeologists recognized these features as part of a food-grinding technology. The data were sorted into groups based on the cultural period of significance assigned by field personnel. Although many sites have multiple occupational episodes, the goal was to distinguish what typical Chaquaqua-style BGS grinding features look like, including the "typical" site assemblage or locality with which they are most often associated, then distinguish the cultural periods to which they most likely belong. Consequently, two categories were created: the earliest period of occupation designated by site recorders and the latest period of occupation. The cultural period of significance was then placed into the existing cultural taxonomy (Zier et al. 1999). The archaeologists' field assessments used radiocarbon dates (of hearths, plants, and bone), rock art panel seriation, and diagnostic artifacts.

Table 7.1. BGS Features in Time (Chronology Adapted from Zier et al. 1999)

CULTURAL TAXONOMY FOR THE ARKANSAS RIVER BASIN		Bedrock ground stone use Characteristics
Paleoindian Stage	**11,500 or older to 7800 BP**	
Pre-Clovis	Older than 11,500 BP	• 6 sites
Clovis Period	11,500 to 10,950 BP	• Open air locations
Folsom Period	10,950 to 10,250 BP	• Exposed bedrock
Plano Period	10,250 to 7800 BP	• Ephemeral use area
Archaic Stage	**7800 to 1850 BP**	• 32 sites
Early Archaic Period	7800 to 5000 BP	• Open air locations
Middle Archaic Period	5000 to 3000 BP	• Rockshelter locations with extensive use
Late Archaic Period	3000 to 1850 BP	• Rock Art Panels (Symbolic)
Late Prehistoric Stage	**1850 to 225 BP**	• 94 sites
Developmental Period	1850 to 900 BP	• Open air locations (increased use)
Diversification Period	900 to 500 BP	• Rockshelter location (increased use)
Apishapa Phase	900 to 500 BP	• Rock Art Panels (Symbolic, Quadrupeds, Hunting)
Sopris Phase	900 to 750 BP	• Rock Art pecked over grinding elements • Tool grooves incised over grinding elements
Protohistoric Period	500 to 225 BP	• No use or decreased use

PALEOINDIAN SITES

The regional Paleoindian culture is known primarily from projectile point technology. The people were highly mobile hunters who probably practiced some foraging, but there is little physical evidence of this behavior. Sites are often temporary camps, although some rockshelters were used during this time. Both the Folsom and Clovis Periods are represented in the region.

Table 7.2. Summary of BGS Features Sites Assessed

	Site Files Assessed (N)	*%*	*Not Assessed (N)*	*%*	*Removed from Sample (N)*	*%*	*Total Sites*	*% All Sites*
Paleoindian	5	0.83	1	0.17	0	0	6	0.01
Archaic	32	0.53	26	0.43	2	0.03	60	0.13
Late Prehistoric	94	0.34	140	0.51	39	0.14	273	0.60
Protohistoric	1	1.00	0	0.00	0	0.00	1	0.00
Undated	40	0.35	75	0.65	0	0.00	115	0.25
Total Sites	**172**	**0.38**	**242**	**0.53**	**41**	**0.09**	**455**	**100**

In the OAHP database, less than 1 percent of the BGS feature sites (N = 6) had a Paleoindian component (Table 7.2). Five sites were available to review for dating methods and accuracy. The BGS feature sites were dated through the presence of diagnostic projectile points. The BGS feature site chronologies begin the Paleoindian (11,000 to 7800 BP) period but have long occupations that end in the Late Prehistoric (1850 to 225 BP). Three of these sites also have diagnostic materials dating to the Archaic (7800 to 1850 BP).

The sites from this time are classified as both open and sheltered sites. However, site records show that the BGS feature is in an open-air context (Figure 7.1). Site 5LA10934 is an example of a typical site. The grinding elements are on exposed areas of bedrock that have been uncovered due to weathering or erosion. The shapes are amorphous. They may have been extensively eroded by weathering processes such as freeze-thaw. The element has no depth, but the surface abrasion is evident from the images. In the case of 5LA10934, an eroding sandstone escarpment nearby might have offered some shelter.

Sites with the Paleoindian to Late Prehistoric components appear to be expedient grinding areas. Many are on surfaces that have fractured due to weathering processes. Although archaeologists recorded their shapes as irregular to oval, none have any depth to speak of, so most grinding event(s) are short-lived, perhaps used for seasonal food processing. However, of the six sites, three sites with an Archaic component (5LA10850, 5LA10934, and 5LA4752) may indicate that grinding occurred later. There is no other evidence from the region that the highly mobile hunter-gatherers of the Paleoindian period used exposed bedrock for food processing. It is more likely that food processing at the sites was related to the later occupations, either Archaic or Late Prehistoric, based on the presence of the stone features at the sites and general trends in the region. However, since we cannot definitively rule out early use of the areas for opportunistic grinding, the inclusion of the Paleoindian component shows at the minimum that BGS feature sites were rare during this time, and they were located in what are today open-air contexts that could have become local landmarks or hunting lookouts.

THE ARCHAIC

The Archaic time frame for the region is divided into three periods: Early (7800 BP to 5000 BP), Middle (5000 BP to 3000 BP), and Late (3000 BP to 1850 BP). The Archaic is characterized locally as a gradual but consistent socio-cultural shift to include more local plant processing by people preferring a still relatively mobile, hunting–based lifestyle. Landscape use became stable as groups of people developed more connections to local territories. The frequency of dates from this time at rockshelters indicates they were transformed into intensively used spaces. The region's population also increased, based on the presence of more sites dated to this period.

Sixty BGS feature sites have an Archaic component, according to the OAHP site files, but only thirty-two of these sites were confirmed as having a BGS feature. The BGS features are primarily associated with diagnostic points. Most of the sites were located in open-air contexts suggesting ephemeral use of exposed bedrock related to temporary encampments (Figure 7.2). However, during the Archaic, we see the first appearance of intensively used BGS features in rockshelters. During the Late Archaic, BGS features became increasingly associated with rock art panels which might have been ritualized spaces (e.g., Classeen 2011; Loendorf 2008). In the JE Canyon sample, BGS features sites of this period are associated with potentially solstice-related rock art, hunting panels, and large maze-like motifs related to meaningful symbolic human expression (Loendorf 2008). These symbolic expressions, coupled with the repetitive motions associated with grinding activities, support the notion of ritual or sacred-use spaces. Metate Cave, although undated, also showcases this type of symbolic rock art in association with the BGS features. Both localities have tool-making elements on or near the BGS features. The tool groove elements at Metate Cave were recycling a displaced BGS feature, while those at 5LA11451 in the JE Canyon survey were created on the rockshelter wall (the exception is a single tool groove element ground over an oval-shaped element). At JE Canyon, the tool panel overlooks the BGS feature. At Metate Cave, the tool groove panel faces away from the main grinding elements. These are exciting patterns that suggest a relationship between toolmaking and resource processing, at least in rockshelter use.

THE LATE PREHISTORIC

The Prehistoric period for the region dated falls between 1850 BP to 225 BP. We see the continued use of bedrock ground stone features along with shifts in both projectile point technology and architecture changes during this period. During the Late Prehistoric, local inhabitants began experimenting with horticultural practices, probably due to influences from both the Southwest and the Southeast culture areas (Stone 1999; Zier et al. 1999). Bedrock ground stone features are found in association with the slab stone architecture (Gunnerson, 1989). Recent research into perishable sandals from Kenton Caves in Oklahoma (an E. B. Renaud site with a BGS feature) and Trinchera Cave suggests that the region was a dynamic human corridor (Rexroth 2010). This evidence should encourage us to reconsider how and when the relatively stable local populations came into contact with these other cultures.

Open-air BGS features continued to be used throughout the region (Figure 7.3). However, we also see more use of rockshelters and the continued use of more intensive BGS feature sites like Metate Cave, Trinchera, and those found at JE Canyon. The trough-like BGS element at Trinchera Cave suggests that grinding practices shifted from oval-like shapes to the more traditional corn grinding shape found in the Southwest. Although we find stone architecture in association with BGS features, the exact nature of the connection is still being explored. Perhaps the stone alignments are related to protection from raiding groups or represent increased storage from the Archaic to the Late Prehistoric. Finally, we must reconsider E. B. Renaud's idea of vertical landscape use in tandem with horizontal landscape use.

Vertical landscapes are focused on the use of resources found in specific side canyons together with those located in nearby floodplains and on the open plains above. The upper terraces may act as boundaries of landscape use. JE Canyon sites are probably an example of this type of landscape use. The notion that the landscape was inhabited vertically brings up questions about how we think about the archaeological sites. For instance, JE Canyon sites are evidently part of a long-term settlement pattern lasting around 500 years but probably much longer. Some BGS feature sites like 5LA11451 and her sister sites, 5LA12781 and 5LA12782 (Lynch 2017a, b), are best accessed along animal trails, whereas 5LA11455 and her sister sites, a lithic scatter above, and a complex rock art panel site (5LA11450) below, could have been assessed from animal trails or by scrambling up the terraces.[1] Regardless, communication between vertically stacked sites is relatively easier than sites that follow the trail system at the canyon's head.

The closest modern example of a lived vertical landscape use is from the Diné peoples of the neighboring Southwest. The landscape use is cyclical as sites are reoccupied regularly throughout the year, from the canyon rims to the floodplains. More than likely, the life history of the canyons was continually linked between the Archaic and Late Prehistoric. The floodplains were places for local animal herds to shelter. The people living in these canyons had already built up generations of detailed knowledge about where the local starches and medicinal plants were to be found. We need more evidence linking the BGS features of the Late Archaic and Late Prehistoric populations in the regions to floodplain farming and incipient horticulture.

Horizontal landscape movement, on the other hand, moves primarily along the upper plains or the floodplains. These areas are more often expansive, allowing for quick and relatively safe travel through open landscapes to other specific localities (Stansbury 1988 (1852)). E. B. Renaud's ideas along these lines imply that the presence of circa 950 BP to 650 BP Apishapa and Sopris phase structures may have been meeting places where the two landscape uses intersected for socio-economic networking or as hostile encounters. While no one has yet tested some of Renaud's more interesting musings, studies using perishables in combination with BGS features could potentially help explore these hypothetical scenarios.

During the Late Prehistoric period, there seems to have been an increase in population movement within the canyons. This statement is based on the multi-layered rock art motifs with shield warriors at Trinchera Cave. Although the introduction of the horse certainly increased mobility through the canyons, that may not have happened until after the BGS features were abandoned. While they were in use, populations through time understood the efficacy of using the hydrology system for trade with groups to the south, east, and far northern plains.

GENERAL PATTERNS

Bedrock ground stone sites throughout the Chaquaqua Plateau illustrate similar trends to other features in the region in that they become more common and more intensively used over time. The main pattern variation in the BGS features throughout the region is the division between open-air sites, those found in open air but near rockshelters, and those found on large boulders within rockshelters (Figures 7.4, 7.5). The intensely used rockshelter sites, like those discussed in the last three chapters, become more frequently used during the Late Archaic and early Late Prehistoric stages. Symbolic and ritual use of these types of sites seems to undergo periodic changes during this time. However, by about 500 years ago, the BGS feature sites appear to have become obsolete. Why?

Bedrock ground stone tools were only one part of a suite of tools. Around 1600 years ago, the people living in the area were beginning to modify how they inhabited the landscape. "Late Archaic materials are gradually replaced by Late Prehistoric materials, suggesting that the local population was undergoing culture change and not being replaced by new migrants." (Campbell 1976:102). By about 950 years ago, the materials culture change had developed into the Apishapa, a culture that may have practiced horticulture (Campbell 1976: 50–53). These changes would have affected how they used the BGS features. Campbell thought the features were still used throughout. Some Apishapa structures were constructed near BGS features (Gunnerson 1989); others like LA211 have materials nearby that can be dated to this period (Campbell 1976:57).

The differences between the open-air sites and those near or within rockshelters might help explore the question. The most parsimonious explanation for the pattern is that there are two distinct types of BGS feature use: expedient and more intensive. The expedient use could be the harvesting of local resources as hunting or processing of game was being done nearby, thus maximizing the capture of food resources for the diet. A second equally plausible notion is that these open-air processing sites represent temporary campsites specifically chosen for the presence of weathered bedrock that could be used to process the local resources as they ripened. An example of this type of use is found in the modern Hadza people of Tanzania (O'Connell et al. 1991). A third possibility is that the open-air sites represent something else, behavioral norms of which we are, as yet, unaware. Perhaps they represent open-air ceremonies or rituals. There is a definite sampling bias of these sites to the military training area surveyed on the PCMS and FCMR, which must be further explored.

The different types of landscape contexts are present during all periods (Figure 7.5), but the number of sites increased dramatically from the Paleoindian period through the Archaic into the Late Prehistoric. The increase could be due to population increase in the region, leading to more BGS surfaces, especially in open-air contexts. Figure 7.5 is based on the number of sites, not the number of features at the sites. There are several grinding features at many of these localities, indicating a long-term campsite or a group of people returning to the same area regularly. The data is fairly incomplete in the OAHP database (less than half of the site files were assessed or had information about chronological and landscape context). Nonetheless, the relationship is an exciting avenue for further study.

Sites like Metate Cave, Trinchera Cave, and those located on JE Canyon Ranch are rare in the region, making up only 15 percent of the sites assessed (26 out of 172). When archaeologists think about "fixed mortars," these are the types of sites that come to mind. BGS features sites in rockshelter tend to suggest several types of use. Those at JE Canyon suggest long-term use of the canyon in various ways—both ritual and large aggregate groups intensively processing food resources (Lynch 2017a, b). Those at Metate Cave support the same type of variation in use but through time rather than through landscape since the physical space within the rockshelter supported many types of activities. These activities are evident through the superpositioning of rock art motifs grinding elements. Trinchera Cave may have seen the same type of variation. However, the BGS features are so disturbed that complete reconstruction of the original, pre-erosional, parent boulder must be completed before the BGS feature elements can be adequately assessed. The presence of overlapping elements and the different ground stone elements suggests different people occupied the space at different times. Trinchera shelter may have supported a stable horticultural community, but it was likely a staging area for those traveling groups moving from or through other areas into the Chaquaqua Plateau.

Three factors regularly recorded on the site forms are distance to water, elevation, and site area. The distance from water to sites increases over time (Figure 7.6). The maximum distance of BGS feature sites to water is much higher in the Late Prehistoric (7,900 meters) than in the Archaic (5,265 meters) or the Paleoindian (2,873 meters). Although some sites are located in areas with access to permanent water, the mean distances are still more than 1,000 meters. BGS feature sites with a Paleoindian component are generally found at the elevations of 4964 feet (range: 4710 to 5125 feet), while those with Archaic artifacts are found around 5073 feet (range: 4400 to 6260). Sites dating from the Late Prehistoric stage have a more extensive range of elevations (4180 feet to 7290 feet) with an average of 5096 feet. The size of BGS feature sites (Figure 7.7) changes significantly through time on the Chaquaqua Plateau. The average size of those sites with a Paleoindian component is more extensive, which is to be expected on sites that have multiple occupations through time. Archaic sites have a greater range in site area than either Paleoindian or Late Prehistoric sites. The caveat here is that many of the BGS feature sites from the Archaic continue to be used into the Late Prehistoric. Those site occupations that begin in the Late Prehistoric tend to be in the open air with a greater range of elevations, further from permanent water sources, and much smaller than other periods. Is climate driving this trend, or could increasing contact with neighboring cultural groups during the Late Prehistoric affect movements across the landscape? Although we can provisionally place bedrock ground stone sites into the regional, chronological sequence, our ability to refine these assessments is severely limited. The information presented here adds comprehensive support for trends Campbell superficially noted in his dissertation fieldwork.

In most cases, sites with BGS features are used continuously over long periods. For instance, the sites with Paleoindian components also have Archaic and Late Prehistoric occupations. While this does not preclude Paleoindian use of the bedrock features, we have to ask whether or not the features date to such an early time. The more critical aspect of this type of time sequencing shows that the sites with BGS features are probably long-term; that is, even those sites in open-air contexts were visited regularly through time. The changing trends in landscape context raise more questions about how the inhabitants of the Chaquaqua Plateau dealt with changing cultural and climate environments.

The continued long-term occupation of some BGS feature sites supports the hypothesis that these sites may act as anchors for many different socio-cultural behaviors simply because they are good places to find water, vegetable and animal resources, and access to shelter. The places are generally accessible to those locals who know where they are. The BGS features are then revisited again and again for any number of social and practical reasons. These data and the continual use of BGS features over long periods seem to support the idea that the landscape was continually occupied by similar peoples until the Late Prehistoric. Afterward, we see increased influence from other cultural areas, followed by the abrupt decline of feature use of about 500 BP. The actual abandonment process probably began earlier, perhaps about 900 BP, but the signature needs to be refined. Sites with Apishapa and Sopris phases near BGS features need to be reassessed to shed further light on the period.

CONCLUSION

Placing BGS features within the regional chronology is going to take much more refined evidence. However, readily available methods such as relative dating and seriation allow us to begin to place the features into likely categories of time. The small step taken here provides the means to test notions about BGS feature variation and use within broader landscape patterns and potentially shedding light on how or whether they were important in the shift to horticulture practices. Relative dating methods, relied upon in this chapter, illustrate the use of BGS features increased through time in both open-air and rockshelter contexts. A Paleoindian component is used cautiously due to the general lack of supporting evidence. It seems more likely that the features were more critical during the Archaic, gaining popularity by the end of the Late Archaic. The Late Prehistoric Stage illustrates the continued intensive use of BGS features throughout the region, especially intense at the rockshelters and increased use of open-air, possibly expedient processing areas. JE Canyon research shows that BGS feature sites with Archaic components remained in use through the Late Prehistoric Stage. We still know little about the abandonment process for these features except that they appear to have been entirely obsolete by about 500 years before the present.

NOTE

1. The sites are within earshot of one another, but liability issues meant that we were forced to take the lengthy animal trails. Had recorders had ropes or a ladder we could have easily scrambled from 5LA11450 to 5LA11455. It struck all of us that anyone living here regularly would have used the more direct vertical option. Indeed, one volunteer spent time trying to find an easy vertical short cut from 5LA11455 to the lithic scatter above. She discovered a small staircase-like path adjacent to the BGS feature we were recording.

8

Transformative Spaces

Creating Social Identity and Reproducing Culture through Grinding Processes

Humans organize their environment for the efficacy of culture production, whether this is the perpetuation of technology or ritual that serves to recreate social identities. Spaces, where cultural knowledge is reproduced are social in nature, being tacitly and explicitly negotiated through time by inhabitants (Godelier 1986; Lefebvre 1991). Due to their expertise in material culture, archaeologists engage with the question of how humans create space to facilitate the reproduction of cultural knowledge because of the implications to sustainable human environments. Although the nature of socially designed prehistoric space stretches the reach of archaeologists because of the symbolic nature of such spaces, building bridges between material culture and the social reproduction of knowledge opens up avenues for constructing sustainable, multivocal visions of the past.

As part of collective work processes bedrock ground stone features are social. They offer unique opportunities to explore how people create spaces to facilitate the reproduction of culture. Social reproduction of space is defined by the practices through which cultural values and norms are passed on from generation to generation through group interaction and the places where social lives are created and lived out (Bourdieu 1998; Lefebvre 1991). A reciprocal situation develops as the spaces embed social behaviors and, in turn, become embedded in social practices (Bourdieu 1998; 1991). Material culture used to produce quotidian needs (for example, food, clothing, and art) becomes embroiled in these socialized spaces (Godelier 1986; Hodder 2012). That is, the material becomes part of lived and imagined experiences that translate into stories, the remains of which have the potential to illustrate how ancient peoples may have reproduced their knowledge systems. Building bridges between the material culture and the stories in which they are embedded, not only contributes to our understanding of how social reproduction works but also enriches our knowledge of change in human social processes. The question is: can, and if so, how can archaeologists meaningfully engage in the pursuit of the signatures of social reproduction processes, and what methodologies enable us to build multi-vocal connections between the material present and past reproduction processes?

This chapter relies on Goffman's frame analysis as a middle-range method to unite multiple accounts (ethnography, colonial encounters, oral tradition, and visual portrayals) of permanent, social food grinding space. The goal is to address the question of how ancient peoples used space as a means of supporting the reproduction of social norms and values. The resulting multi-vocal perspective is then applied to the problem of bedrock ground stone features found along the Chaquaqua Plateau. However, the importance of modeling bedrock ground stone features has been illustrated in previous chapters. The process explored here provides insight into the human systems behind the phenomena observed in the field, providing a means to support archaeological interpretation.

ORAL TRADITION AND MATERIAL CULTURE

Language structures human social and environmental landscapes, affecting how we think about and navigate through those realms. People imagine and then talk about the world in which they participate, re-imaging material culture within familiar cognitive structures. Ideas are developed, negotiated, and sustained within the language contexts that reflect how we understand the world and make sense of events unfolding around us (Bakhtin 1981:8–16; Goffman 1974; Lévi-Strauss 1969). How people think about, and use language is often "haunted" by the physical landscape (Basso 1996). Epic adventures, mythic chronicles, and personal accounts create "text artifacts" (see Silverstein 1996) that can be used to construct a social context for archaeological material culture, helpful in exploring multiple perspectives or frames. The context within stories frames the audience's perceptions; thus, when grinding episodes are depicted, the spaces in which they are placed can be used to interpret extant archaeological spaces. At the same time, a cultural preference for living in space frames how participants and observers talk about space or activities occurring there. The process creates a storied landscape where people engage with material culture in meaningful ways. Archaeologists can use these text objects to formulate questions about the archaeological material culture, thus providing additional lenses to view ancient landscapes and reduce observer bias in terms of questions posed and data interpretation. These lenses provide frames of reference to the material world.

Frame analysis requires multiple perspectives of a single event from which behavior can be broken down into meaningful, analytic units. Events related to the social reproduction of space at permanent ground stone features include quotidian production of food resources, ritual food creation (first fruits ceremonials), or food manufacture within the context of rites of passage (for instance, puberty rituals, marriage). Material culture (grinding tools) and space where they were used (bedrock ground stone features) can then be analyzed as part of a social production process. Frame analysis is a flexible method for text analysis that incorporates as many perspectives as are available while being able to add more texts as they become known to build an ever-evolving view of grinding practices and associated material culture.

The texts evaluated for this chapter include Native American oral tradition, drawn from the western and southwestern United States, historical narratives and images, along with ethnographic writing and images that are centered on grinding activities at permanent ground stone features (BGS features, but also fixed metates such as the Southwest mealing bins). In examining ethnographic and ethnohistoric narratives from 24 cultural groups, there are more than 455 narratives, including transcriptions and translations of orally transmitted tradition. Twenty-two provide a direct reference to staple food processing at stationary grinding features, although there were quite a few more references to food production (Table 8.1). Three ethnographic accounts showing slightly different views of grinding space were selected for comparison with two Euro-American accounts and four grinding stories from the California and Pueblo oral traditions. Juxtaposed with these written records are visual images of grinding spaces: two photographs from the southwest (Figures 8.1, 8.2), one photograph and a wood carving from acorn grinding cultures (Figures 8.3, 8.4) and a discussion of two Hopi *katcina* drawings collected by Jesse Fewkes (1903) from the Zuni Pueblo.

Ethnographic Frames

Edward Gifford, an early and prolific recorder of native California cultures, published ethnographies and descriptions of social organization, as well as oral traditions (Barrett and Gifford 1933; Gifford 1916; 1917; 1923; 1932). His interest in acorn specialization provided some of the earliest

Table 8.1. Summary of Food Processing Oral Tradition Grinding Space

People	*Source*	*Narrative*	*Food Reference*
Chukchansi Yokut	(Rogers and Gayton 1944: 195–196, 202)	Coyote Makes Mortars Holes and Pestle	Origin of Acorn Bread
Gashowa Yokut	(Kroeber 1907: 209)	Coyote, the Hawk and the Condor	Acorn Mush, Deer Meat
Gitanemuk	(Kroeber 1907: 243)	The Panthers Children and Coyote	Acorn Mush
Hopi	(Titiev 1948: 37)	The Story of Muyingwa	Corn for grinding, piki
Hupa	(Goddard 1904: 126, 147, 183)	Creator and Culture Hero	Acorn Meal
		Dug from the Ground	Gaering Acorns
		Goosebury Place Brush Dance	
			Acorn Soup
Rumsien Costanoan	(Kroeber 1907: 200–201)	Coyote	Acorn Mush, Acorn Bread
Wappo	(Radin 1926: 50)	Bear Woman and her Children	Pick and eat Acorns
Western Mono	(Gifford 1923: 346, 355, 364)	Prairie Falcon Adventure	Acorn Bread
		Coyote Decides to Catch Gophers	Acorn Soup and Deer Meat
		How Gull Became Deer	Acorn Bread
Wintu	(Du Bois and Demetracopoulou 1931: 290, 299)	He Who is Above	Origin of Acorn, Acorn Bread
Yana	(Sapir and Dixon 1910: 88, 103, 133)	Origin of Sex and Hands	Acorn Mush
		Coyote and His Sister	Pounding Acorns
		Coyote Pine Marten and Loon	Acorn Bread; Acorn Soup
		Woodpecker and Woodrat	Acorn Mush
Zuni	(Parsons 1931: 140, 141)	Kuchininaku chase butterflies	Grinding
		Bewitched into Coyote	Ground Corn

ethnographic descriptions of acorn processing at bedrock mortars (Gifford, 1936). Gifford writes that wheat was pounded

> in a mortar hole six or seven inches deep. This was filled nearly to the brim with slightly parched wheat. At first the miller struck it gently with the 10-pound pestle, which, grasped with both hands, she raised only 2 or 3 inches above the wheat; then as the grain was gradually broken, she raised the pestle higher and higher, striking harder and harder. . . . The miller sat with her left foot under her right buttocks, her right foot folded at her side; at times she sat crosslegged (Gifford 1932:24).

Interested in how the Western Mono utilize their technology, Gifford provides a straightforward account of the miller (a woman) at the workstation. Information provided is the gender of the worker (female), the type of food being pounded (wheat), the process of pounding (steadily and more intensely as the work goes on), and the miller's position (usually sitting) while she works. The frame of reference centers on the grinding space, thus providing the archaeologist with crucial information about the functional aspects of material culture. However, the description lacks information on the ritual or quotidian nature of the grinding event; we are not provided with the spatial organization or symbolic significance. Later, Gifford (1932) adds that millers gather together at the workspace, as much to process food as to communicate with each other, thus allowing the archaeologist to infer that communal food-processing sites create opportunities for social solidarity through small talk and perhaps gossip, which is a vital part of the production (for example, through information transfer, meaningful negotiation, resolution of conflicting information) of knowledge.

In comparison to Gifford's account, A. H. Gayton's (1948a, 1948b) ethnography of the Yokut and Mono peoples provides a view into household relationships when describing milling spaces and the activities that surround them. Such relationships, at the most basic level, consist of culturally constructed social positions of individual members to one another, the values they share, the duties they owe to the household unit, and how individuals within the household space were expected to behave or the limits to which an individual's behavior could be negotiated.

Gayton (1948b:178) describes the environment surrounding the social preparation of newly harvested acorns for drying. Every member of the household has a role to play, including entertainment, "During the evening young people shelled acorns while some old person told stories or adventures" (Gayton 1948b:225). Gayton does not reveal the subjects of the stories. However, the inclusion of "some old person's" knowledge as a diversion, or possibly reward, during food preparation offers insight into how acorn processing and material culture can encode social values and norms, but more importantly, how generational knowledge is transferred generationally during communal food processing. Gayton's readers are presented with an intimate view of the setting, along with a lengthy description of the grinding space itself,

> The fine skins of these (acorns) were taken to the bedrock mortar. As the woman pounded up acorns in the holes, she took the flour out from time to time, piling it up beside the hole. When she had a sufficient quantity, she took up a handful, put it on a circular, flat, winnowing tray. The meal on the tray was then shaken in such a way that the coarse particles came to the surface, rolled off onto a similar tray. The fine flour stuck to the basket surface. The finest flour was set aside, the coarser returned for further pounding. This routine was repeated about three times with the same batch of flour. Nearly a whole day was spent at this work, producing a quantity sufficient to last an average family two to three days. . . . At the end, there should not be more than a handful of coarse meal, remaining in the mortar hole; this was discarded (Gayton 1948b:178).

Gayton's observations provide insight into quotidian social production through food processing from the acorn to the final household food supply. Tedious work (shelling) is attended by an elder's storytelling. Cultural production is found in the storytelling (for example, values and norms) and the shared labor (knowledge of how the staple food is to be processed, along with the expectation that all members of the household are expected to work). The storytelling (traditional knowledge) or gossip (group information) can be viewed as part of a reciprocal exchange that is concentrated within the intersection of the workspace. The labor, social environment, and work processes encode a worker's role as provider for the family's staple food, the value placed on creating "the finest flour," and social identity reinforced through the process of grinding (milling) through stories, gossip, or other mechanisms. The themes of craftsmanship as social responsibility are comparable with other accounts provided below.

Historical Frames

Spanish historical descriptions provide the earliest Euro-American chronicles of grinding spaces in the Southwest. Although more concerned with the pursuit of wealth, the Spanish often recorded details of grinding events observed on expeditions north of the Rio Grande (Table 8.2). The Spanish accounts illustrate grinding activities taking place within households at fixed grinding spaces called "mealing" bins, providing glimpses into the social processes surrounding the events. *The Relacion de la Jornada de Cibola, Pedro de Castaneda de Najera's Narrative, 1560s (copy 1596)*

Table 8.2. Euro-American Accounts of Stationary Food Processing

Spanish Narrator	*Year*	*Technology*	*Process*	*Food*	*Source: page*
Casteno de Sosa's	1591	2–3 grinding stones	flour processed	bread, flour, gruel, tamales	(Hammond and Rey 1966: 277)
Antonio de Espejo	1582	5–6 women together	process together at a single mill	Breads	(Hammond and Rey 1966: 219)
Luxan	1582	4–8 grinding stones	fine flour	"very fine tortillas"	(Hammond and Rey 1966: 172)
Chamuscado Rodriquez Expedition	1580s	stationary millstones	corn and other foods	daughters grind	(Hammond and Rey 1966: 85)
Coronado	1540	"best arrangement and method for grinding"		"best tortillas"	(Hammond and Rey 1940: 172)

describes a food processing scenario encountered near the Albuquerque area,

> They have houses that are excellently divided up [and] very clean, where they cook food and where they grind flour. The [latter place] is a separate room or small secluded room where they have a large grinding bin with three stones set in mortar, where three women go, each to her stone. One of them breaks the grain, the next grinds [it], and the next grinds [it] again. Before [the women] go through the door, they remove their shoes and gather up their hair. They shake their clothes and cover their heads. While they grind, a man is seated at the door playing [music] with a flute. To the melody they draw their stones and sing in three parts. On a single occasion they grind a large quantity [of flour], because they make all their bread, like wafers, from flour mixed with hot water (Flint and Flint 2005:419).

Najera's description provides a glimpse into the process of cooperative labor necessary to produce high-quality, high-quantity amounts of flour for group consumption, although it is unclear whether the flour is for quotidian or ceremonial use. The context, however, is a social one. Labor is organized at the grinding stones (stationary, separate from the other household workspace). The communal work is supported by music, provided by a male flute player (presumably a household member, but perhaps just someone wishing to invest their own labor in the food-making process, as is the situation related by Parsons 1930). Social reproduction, in this case, takes the form of maintenance and creation of social identity as inherent in food processing events as evidenced by "each to her own stone," or the man's identity as a musician whose song is intimately intertwined with the grinding process, all parts (flute music, grinding to the melody, and singing that is harmonized with both illustrate the synergy of multiple social identities).

Luxan's account of the Espejo Expedition resembles Gifford's formal account of grinding processes. He gives the following description of the milling processes he observed

> at the end of these they have their mills where they grind the corn for the making of tortillas. They grind it raw or toasted and knead it with hot water; they make very fine tortillas. Their mills are of the following type: four or five or even eight mills are placed together depending on the capacity of the house. They are a yard long and two-thirds wide they are made of white-washed stones, built low, right on the ground, and resemble metates . . . the Indians grind with another stone (Hammond and Rey 1966:172).

In Luxan's description, the miller is an extension of the grinding technology. While de Najera spends time describing the millers, their action before the milling event, and their place within the context of the social space. Luxan's report focuses on the number of grinding stones fixed in the space. The social context surrounding the milling event and even the gender of the grinder is completely lost in the technology. However, the reader is left with a sense of repetitious behavior at the foundation of the seemingly ordinary event. Contrast Luxan's description with an account given by Crow-wing in the *Pueblo Indian Journal* of the preparation, social relationships, and duties inherent in preparation for a marriage ceremony.

> The girl grinds corn all day. Some of the Rabbit clanswomen go and help the girl grind. Early in the morning the girl gets up to do her work. All day long they are busy in the house—some women are cleaning house, some carrying water; and some boys go for wood. . . . This is the last day for the marry girl to grind, so now everybody is busy. The women are making *piki* to take it to the house where the girl is. The boy's father's brothers go and kill sheep, so I [being a Bear clansman] go and kill my own sheep to help the boy. Also we buy some goods to take to the boy's house. When evening comes the girl's folks take corn meal to the boy's house, and some *piki*, too. Besides, the women and girls go there and make *piki* all night. The girls and women have to stay up all night with the girl (Crow-wing and Parsons 1925:35–36).

As a member of a close-knit community, Crow-wing provides the pertinent information about social obligations (e.g., Rabbit Clan—women; Bear Clan—men) and the expectations (everyone is busy) surrounding a grinding event related to a rite of passage (girl's pre-marriage grinding for community and in-laws). She has been grinding within the ceremonial context for days in preparation for her upcoming marriage ceremony. The community is interested in the girl's progress, and women help with the ritual preparation of the house and ensuring food and water are plentiful. Technology (including grinding space) is subsumed to ritual and social obligations at the heart of the social reproduction of grinder's identity as central to the community's creation. Every act surrounding the grinding event imagines the transformative process being performed by the grinder; while she works away grinding corn, her identity is also being ground into something new. The "marry girl" grinding is unleashing the potential wife and mother. All in the community are well aware of and support the ritual change in social identity.

Historical narratives offer a perspective on the nature of grinding spaces but only hint at the importance of the grinding space to the reproduction of social life and cultural identity. However, the material culture is deeply embedded in ceremonial rites of passage as well as within kinship obligations. Crow-wing's observations, though lacking the description of the material culture, provide clues about the social values and norms that are part of the process of grinding. Although the actual workspace, where the grinder performs her ritualized task, is hidden from the public, the audience is brought into her presence by Crow-wing's observations. The community understands what is at stake. The significance of the grinder's work in recreating the fabric of society is not lost.

Corn Grinding Narrative and Imagery

Stories from the oral tradition and the images drawn by Native participants illustrate the social side of material culture. The space in which grinding activities occur is shaped by the symbolic and meaningful frames of the participants. The audience is moved ever closer to the social reality experienced by grinders.

The narrative landscape of permanent grinding spaces in the Southwest, viewed from a participant's frame, merges public and domestic communal space, often consisting of multiple women processing food for community consumption. Groups of grinders work in private and public spaces amid the watchful eyes of more experienced women (aunts, mothers, mothers-in-law, or grandmothers) or are entertained by groups of men (Beaglehole and Beaglehole 1937:31; Frisbie 1967:35–37; Green 1990:203–205; Sekaquaptewa and Udall 1969:110–111). The act of grinding and food preparation is a social gathering but also serves to reinforce cultural values and norms (Wall and Masayesva 2004:450). The reproduction of values is buttressed by the stories incorporating the imagery of the grinding event.

The Hopi story *The Hicanavaiya Myth* (Titiev 1948:32–37) provides an example of the social complexity embodied within grinding events. The heroine, Pavinmana, a young girl from Oraibi, is abducted from her village by her pet water antelope (a *katcina* named Hicanavaiya). He steals her away to his village in the San Francisco Mountains to undergo an "ice ordeal" from which few girls survive. She is kept isolated from the rest of the community until one day,

> Pavinmana was thereupon led from that room and fed well. When she was through with her meal Hahai'i said, "Now I'll let you grind corn. . . . If you have the right power you may save yourself" . . . she brought out great chunks of ice that she put on a metate made of ice for Pavinmana to grind . . . Spider Woman directed Pavinmana to work steadily because they would win out if she could grind up all the ice without getting tired. Thus encouraged, Pavinmana worked steadily on. As the metate filled with water Spider Woman dipped it out and poured it onto a container, and the girl continued at her task until all the ice, including the metate, had melted. Then Spider Woman said: "Thank you, we have come out all right. . . . Soon you will get real corn to grind. Remember, treat these people in a kindly way." (Titiev 1948:34)

In this section of the story, the audience is brought into the room with the grinder. It resembles Crow-wing's ethnohistoric description but contains symbolic structures (Spider Woman, ice grinding) that allow the audience to imagine the experience or re-connect with their own rite of passage.

The grinding space in which the young girl must prove her skill and endurance is set apart from the rest of Hicanavaiya's community. Pavinmana is among strangers and alone, except for the presence of Spider Woman. Although Hicanavaiya's in-laws cannot see the ordeal going on behind closed doors, they too can imagine events. Those who may have experienced a similar grinding ordeal for themselves may re-live the challenge (e.g., van Gennep 1909; Whitehouse 1996). The ritual process suggests that the girl's experience is reinforced by separation from her previous identity and natal community. Her new social role as a married woman can only be assumed if she passes the grinding test.

In the imagined space of the narrative, the ice ordeal seems impossible to accomplish; it is painful and induces fear of death, if not real death, then certainly a social one if the initiate fails. The grinding ordeal is in effect creating a new social person. Although it is a difficult task, the girl has the power within her to succeed if only she does not give up. In the narrative, the familiar food processing tools are transformed from ordinary stone to ice; the familiar corn also becomes ice. Surviving the ordeal of grinding ice on ice is crucial to the reproduction of social knowledge, creating one's social identity and, in this instance, creating solidarity with future kinsfolk. In the end, even the ice metate is consumed by the ritual transformation.

Observers of the ritual are kept outside of the grinding space, but as we are told elsewhere in the story, they seem to be fully aware of the girl's progress. The girl is not entirely alone since Spider Woman, a powerful figure in the Hopi belief system, is watching over her. Spider Woman also encourages the girl as she progresses through the ritual stages by removing water from the grinding area. Spider Woman's support can be compared to the "Rabbit clanswomen" who support the "marry-girl" in Crow-wing's account. In the Hicanavaiya story, the girl, since she pays attention to her guide, succeeds in a task at which many others fail. It is within this narrative landscape that the archaeologist has the opportunity to observe the social complexity of an individual's place and responsibilities within the broader community in a functionally complex space such as bedrock ground stone features.

The anchoring of social transformation processes and the creation of social identity to grinding spaces are also found in the Zuni tradition *Young Women and Old* (Parsons 1930:19–24). The tale was collected in 1918 from a member of the *Ne'wekwe* clown society (Parsons 1996:740). The first half of the story concerns the transformation of two elderly women into young girls through the actions of *Aihayuta* (spirits). These old women-young girls seduce the community's most eligible bachelor. A grinding event placed at the center of the story recounts the girls grinding but doing so to the flute music of a neighbor boy (the poorest wretch in the pueblo). The second half of the story concerns the social transformation of the most eligible bachelor into an insignificant character (because he rejects the old women-young girls when he finds out they are not young girls) and the poor wretch, neighbor boy into the most influential person possible: a rain priest.

The impact of this story for archaeological inquiry is the anchoring effect of the grinding space in the creation and destruction of social identity, an essential aspect of social reproduction and the maintenance of social space. The grinding scene occurs, in the old women-young girls' house, after the handsome bachelor rejects them (Parsons 1930:21–22). The scene seems to be an ordinary domestic world of grinding food for daily consumption when the poor boy from next door begins to play music and sing for the women while they are working. The grinding description, in some ways, mirrors that of de Najera's and other ethnographic accounts of Pueblo domestic grinding but pulls the audience into the workspace as sympathetic participants. Even the most ordinary material culture seems to possess the ability to anchor and create a space for social movement. Oral tradition, with material culture descriptions, offers valuable insight for framing research questions but also for the interpretation of the past.

Native and Ethnographic Visualization of Grinding Space

Hopi *katcina* drawings at Zuni Pueblo support the intimacy of grinding events within the fabric of society. Hopi artist's renditions of the grinding practices are embodied in two images of *katcinas* taking part in the *Palülükañti* or *Añkwañti* ceremony held in March (Fewkes 1903: 24, 26, 42). The first drawing is a marionette sometimes used in ceremonials (Fewkes 1903: Plate XXVII). The Corn Maidens are kneeling before a mealing bin adorned with two birds along the top of the bin. In the marionette, the birds move back and forth, and the maidens' hands can be moved to imitate real-world grinding. Along the top of the mealing bin is a row of rain cloud symbols that is mirrored on the bottom of the image.

The Corn Maidens wear a headdress symbolizing the raincloud. Two symbols run horizontally on either side of the headdress, while two symbols run vertically down the center. An ear of corn with feathers on either end is painted across the maidens' foreheads; one eye is blue, the other is yellow. The maidens have turquoise earrings with coral necklaces. Their dress is a blanket draped around an underskirt made of feathers with symbols of rain clouds and falling rain. The names given the figures are Hopi *Calako Mana* or *Palahiko Mana* (Fewkes 1903: Plate LVI).

The second image is of the reenactment of the grinding portion of the *Palülükañti* or *Añkwañti* ceremony. In this image, two *Añya Mana* actively grind corn as part of a yearly ceremony (Fewkes 1903: Plate XXXII). Behind them are four *Añya Katcinas* who sing for them as they work. At either end of the mealing bin are two *Hehea* (also *He'he'a*) *katcina*s with zigzag symbols on their faces, diagonal eyes, and "phallic symbols" drawn on their arms and legs. The *Hehea* carry in the grinding stones and setting them before the fire for the girls to work (Fewkes 1903: 46). Across from the grinders is the *Paiakyamû* or "Hano glutton" (Fewkes 1903: 178; Plate LVIII). The kiva chief and the *Hehea* converse about how "the girls are clever meal grinders." The ceremonial is performed both privately within the kiva and for the public.

Images collected from Hopi artists (Fewkes 1903) contain extremely vivid spatial imagery of grinding loci from a distinctly Puebloan worldview. Ordinary grinding tasks are performed by Corn Maiden *katcinas*, ancestral cloud beings, who act as connections between the extant, experiential world and the ancestral one (Stevenson 1904). The Corn Maidens dominate the grinding space yet are intimately connected with the community. The images seem to suggest a space where ancestral grinders (*katcinas*) meet real-world grinders (the audience), providing a continually renewed connection between generations of grinders. The image illustrates one frame of a traditional view of Puebloan grinding space as part of a social world linked with that of the ancestors (see also Crow-wing and Parsons 1925: fig. 7).

Contrast the description of these images with those of Curtis (Figure 8.1), Mooney (Figure 8.2), and acorn-pounding photographs (Figures 8.3, 8.4) collected around the same time in California. The images demonstrate the quotidian side of Puebloan milling activity, including ordinary clothing and locality (within and outside the home). The images capture the technology of grinding with associated baskets, cooking tools, and food-making processes. These images were not intended to convey the symbolic framework represented by grinding. The production of social space from the ethnographic frame is contained within the material culture but remains somewhat static. It is the Hopi artist who provides the production process: annual ceremonials acted within the private and public domains that symbolically turn every grinding act, no matter how ordinary, into an experiential, representative landscape of the Corn Maidens, *Anya Katcinas*, *Hehea*, and *Paiakyamû*. Even the most ordinary grinding activity is not a static space but one where the "clever" grinder could engage with the world of the ancestors, her community and hear Spider Woman's encouragement simultaneously.

Acorn Grinding Narrative and Visualization of Grinding Space

For cultures that relied on the acorn harvest as a dietary staple, the processing events traditionally took place on bedrock outcrops near villages, households, or acorn groves. Small structures were sometimes constructed over the bedrock ground stone features to protect grinders and the food being produced (Barrett and Gifford 1933:27; Gayton 1948a:62; Powers and Powell 1877). Women often processed acorns in groups during the harvest season, during festivals (for which portable mortars may have been used), and as needed throughout the year (Barrett and Gifford 1933:143, 208). Aside from exchanging information, bedrock mortar sites were spatial hosts to various social production processes, including proper food preparation, relationships between household members, and reproduction of cultural knowledge.

Bear and fawn stories are perhaps one of the most widespread stories within the California cultural tradition (Dangel 1929; Du Bois and Demetracopoulou 1931; Gifford 1923; Goddard 1906; Heizer 1978; Kroeber 1909; Powers 1882; Radin 1926; Rogers and Gayton 1944; Sapir and Dixon 1910). While not all of the stories include bedrock grinding features, many of the stories encode the production and maintenance of proper social relationships and the value associated with them, including kinship responsibility, gender power structures, and maternal devotion. One version of the story recorded by Gifford (1923) provides an example of the social complexity embedded in grinding spaces and material culture. In this story, the bedrock ground stone space offers a haven for two fawns who have avenged the death of their mother, Deer. Deer was killed and eaten by Bear (her sister-in-law) while they were out gathering at Bear's gathering place. The fawns find their retribution by killing Bear's cubs but then must escape their aunt's rage. Gifford's consultants tell how the fawns

> traveled all over the mountains, looking for a hole whereby they could enter the underworld. At last, after a vain search, one said, "I think we must go back to where our mother pounded acorns. There we shall be able to enter the underworld." "All right," the other brother replied. They went to that hole in the granite, where their mother had pounded acorns. They entered the underworld through the mortar hole and closed the opening with the pestle. They instructed the pestle, "When bear comes here, hit her as hard as you can." (Gifford 1923:357–359)

The bedrock ground stone space offers the only possible escape route for the young fawns from their vengeful Aunt. The mortar and pestle, possessions of their mother, the grinder, remain imbued with her social identity as a mother (nurturer) and provider (protector) for her family even after her death. For archaeologists, the point to remember is that the bedrock ground stone feature retains the power and abilities of their owners, thus possessing a life history that persists after abandonment. The visualization of mortars as portals to an ancestral world indicates the importance space holds in this cultural frame. The narrator of this story incorporates the bedrock ground stone space as a transition between this world and the "other," between an in-law and natal relationships, linking the material reality with a supernatural world, thus providing a glimpse into a space that intimately connects the production of cultural values and norms.

The final narrative of grinding spaces explored here is an excerpt from *Pine Squirrel and Frog Woman*, a Wintu tale (DuBois and Demetracopoulou 1931). The story is a long narrative divided into two halves that center on an acorn milling event. A young girl seeks out the handsome and eligible Pine Squirrel, who lives with his parents. At the end of three days, she goes out to pound acorns with Pine Squirrel's mother,

> One day she was going to pound acorns. Old man went for wood because he liked her. Young fellow went out to hunt. She took her hopper and set it on a rock and pounded bush acorns. She pounded and pounded and pounded. She put her meal in a shallow basket and the old woman helped her sift it. The girl pounded, and the woman sifted. The coarse part of the meal dropped down. The girl heard it fall and the old woman put the coarse part back in the hopper. The girl got up and went home. She was ashamed not to have pounded the acorns fine enough (DuBois and Demetracopoulou 1931: 331–332).

Pine Squirrel eventually locates his betrothed, but she refuses to return with him because of her inept grinding skill. This story echoes the ethnographic depiction provided by Gayton (1948a, 1948b) and *The Hicanavaiya Myth* recorded by Titiev (1948). The portion of the tale that occurs at the acorn processing space provides context for the grinder's social identity as a provider for her family. The reader is furnished with the technological process and material culture, but this time the narrative is told from the personal perspectives of the grinder and her in-laws. The grinding space both links and, in this case, destroys the dreamed of marriage alliance. The grinding ordeal, unlike that in *The Hicanavaiya*, was a catastrophe and the grinder, perhaps more than anyone other character in the story, has the responsibility (and power) to uphold the social consequences of this self-recognized maladroitness.

The narrative landscape of bedrock ground stone features reveals a space where social bonds are developed. One learns about their own role within society, and social identity as a grown person within the broader community is nurtured. Each story discussed presents a different frame of the "real" grinding place, yet all have elements depicting the production of social space. The loci immediately surrounding the grinding event enclose the heart of the nuclear family unit, where social relationships are first explored. Mothers grind food with children by their side. Young girls grind together. Young girls pound while their mothers-in-law winnow flour. Social identities, forged or destroyed at the ground stone feature, permeate the community and the lived experiences of the grinders.

Archaeology and Permanent Ground Stone Space

The archaeological contexts for bedrock ground stone features used in the acorn production process are well-documented in North America, as are the Puebloan fixed mealing bins associated with corn production. Both cultures have household and community grinding spaces considered to be the domain of women, including ritual and labor organization, but the reproduction of culture at these workspaces was not exclusive (Bartlett 1933; McCarthy et al. 1985). Men are present. Their participation is primarily that of a support role, "Although mortar stations were used exclusively by women, information about women's activities as well as men's activities was exchanged at these places" (Jackson 1991:317). Children are present. They support and imitate the grinder or find protection in their mother's workspace. As the anchor of community presence, knowledge, and activities, the spaces come to "reflect the social organization" (Ardener 1981:12) but also echo the social beings (living and ancestral) who regularly engaged with the repetitive activities enacted within that space.

The inclusion of symbolic frameworks of grinding social space in both Pueblo (corn) and California (acorn) narratives belies the differences in the construction of respective grinding spaces. Southwestern, horticulturist spaces are constructed within households or outside permanent structures primarily for communal processing of food resources. The spatial patterns vary but generally consist of constructed mealing bins with quarried metates (ground stone surfaces) permanently fixed within the bins (Bartlett 1933; Beaglehole and Beaglehole 1937; Eddy 1964). Varieties of cultivated corn were harvested, stored, then ground at the mealing bins. The spatial organization facilitates the group processing of flour. The milling spaces of the Southwest are constructed for millers to work side by side, whereas acorn milling spaces are positioned across a rock surface and may have allowed more flexibility in the individual position of the workers through fluid space.

Acorn ground stone features were constructed on naturally occurring outcrops of granite bedrock near encampments and the tree groves where acorns were harvested (Bennyhoff 1956; Powers and Powell 1877). Milling space variability ranged from intensive acorn processing areas composed of large, exposed bedrock boulders containing clusters of milling surfaces to multi-purpose processing sites with individual surfaces varying in size, shape, and function (McCarthy et al. 1985; True 1993; True et al. 1991). Variations in the site pattern range from exposed bedrock with few clustered mortars to sites with rock art and incised lines near and on the milling boulders (True et al. 1974).

Acorn milling cultures were primarily mobile societies with semi-permanent winter villages, whereas the southwestern corn-grinding cultures lived in permanent villages and were corn-growing specialists. As part of seasonal procurement strategies, acorn gathering groups moved from sedentary winter villages to temporary fall encampments near tree groves harvesting other resources as they moved through the landscape (Bennyhoff 1956; Haney 1992; Parkman 1994; True 1993). Group harvest strategies varied according to the distribution of the resources across the landscape, but Jackson (1991:301) suggests "women's food procurement and production activities were fundamental in structuring subsistence and settlement systems and other inter-and intra-community social and economic relations."

FRAMING GRINDING NARRATIVES TO THE CHAQUAQUA PLATEAU

The bedrock ground stone features along the Chaquaqua Plateau present a daunting archaeological problem. There is no ethnographic corroboration for the importance of the BGS features to cultures that inhabited this region of the High Plains. As we have seen in previous chapters, temporal control is associated with surrounding features and artifacts such as hearths, rock art panels, structures, or diagnostic projectile points. Since no record of the use of these features has been found, it appears that the bedrock grinding areas were no longer in use by the time Athapaskan speakers appeared in the region about 400 to 600 years ago.

The appearance of bedrock ground stone features in southeastern Colorado has been primarily addressed in terms of changing food procurement strategies: either a move to corn horticulture or reliance on local plant resources (see discussion in Lynch 2017a). However, the inability to date these features directly or determine the nature of the resources being ground further exacerbates the archaeological situation. Moreover, when encountered in the field, the common response to these features is to assume a generic vegetal food processing scenario and call it good.

The case study presented here uses frame analysis to consider the Colorado BGS features, as material culture intimately entwined with past social reproduction processes. It asks how prehistoric peoples of the Chaquaqua Plateau socialized this landscape by creating permanent ground stone features. At the heart of the question is how we can better understand BGS features as part of the broader, regional archaeological material culture. The data presented so far in this book has left many unanswered questions about BGS features. Are the grinders entangled groups of related family members gathering to work together to process local food resources? Do grinding spaces, such as JE Canyon, Metate Cave, or Trinchera Cave, represent small households or indicate larger aggregations of people? How are the open-air sites related to the more intensively used sites? How do the archaeological spaces facilitate social reproduction or help us understand past social identity?

Defining the space as part of social reproduction processes requires a framework to develop questions about the specific features and the organization of the ground surfaces on these features. It may not be enough to assume if or how the surfaces are related to one another. Field recording procedures and 3D modeling refine ground stone topography to address whether the surfaces could have been used at the same time or if they reflect multiple events with a single grinder (Lynch 2017a, b; Lynch et al. 2017). However, the context in which children or multiple adults might have used the BGS features is required to address questions of social reproduction processes. Lynch et al. (2017) used 3D modeling to determine the position of millers for group grinding and information exchange through the orientation of ground stone surfaces.

SOCIAL REPRODUCTION OF GROUND STONE SPACE

Reconstructing the social world of prehistoric grinding is heavily reliant on living cultural models, but the framing matrix presented here illustrates how grinding activities reinforced a variety of social mechanisms, thus providing the means of inferring reproduction of social space (Table 8.3). The frames focus the grinding space, regardless of regional variation in site arrangement or resource processing, squarely on the community (and associated material culture) created by clustering individual grinding practices at meaningful places. Bedrock ground stone sites offer a "social map" (*sensu* Ardener 1981:11–34) of grinding practice radiating from the work surface to the worker; embracing the innate expertise (for example, physical strength, imagination, and capabilities), equipment (ground stone surface and hand tools), access to knowledge and resources (for example, household unit, community support, extended family), and relationships with the resource being ground (landscape knowledge or tenure). The commonality of references to the staple foods, recipes, and food processing within oral tradition from two seemingly discreet traditions illustrates the cultural food practices. The central role of food processing within the cultural context suggests the person(s) processing food resources possessed a skilled understanding of how to manipulate those resources in normative ways. The social identities of the participants radiate from the ground stone feature. They are the social space of grinding where social identity is negotiated and maintained within the worker's family and community. The spatial organization of the material culture at the heart of the grinding space, is visible in the archaeological record.

A simple illustration to map this interactive space places the grinder(s) at the center of the social world held steady in the social space by the material culture (Figure 8.5). The permanent ground stone space draws the community to the grinder's workspace. Surrounding the worker at the most interactive level are her household members, children, parents, grandparents, and in-laws. Young children still dependent on the protection and nurturing of elders learn about their place in the family and broader community. The feature is the intimate knowledge arena where the bedrock ground stone space becomes framed by broader social values, and "behavior and space are mutually dependent" (Ardener 1981:12). Bedrock ground stone places represent the materiality of this social construction.

By simplifying the key aspects of social life anchored to ground stone space, a matrix of social reproduction processes can be illustrated (Table 8.3). In this study, there are eight resulting frames of reference for events taking place at permanent ground stone spaces (ethnographic accounts, Spanish accounts, Crow-wing's accounts, corn and acorn narratives, ethnographic images, Hopi images, and archaeological research). Some frames are from participatory perspectives, while others are observations. However, all accounts offer support for how the reproduction of knowledge takes place. Social information is communicated through activities, thinking about activities, and relating that knowledge through stories, symbolic thought, and ritual practices. Spatial information is sometimes abstract, for example, "portals to the underworld." However, it also shows ground stone technology in the lived-in world and connected to the world of the ancestors.

Creating a Multivocal Structure of Bedrock Ground Stone Spaces

The frames above provide a starting point for formulating hypotheses for observations made of bedrock ground stone features applicable to grinding spaces on a broad scale. The incorporation of BGS feature spaces into social frameworks, using two different subsistence traditions in North America, offers support that grinding cultural practices are more dependent on social processes occurring with space rather than on the type of

Table 8.3. Summary of Frame Analysis

Frames	*Insider Perspective (being, doing)*	*Outsider Perspective (fact-based observation)*	*Social Information Captured*	*Spatial Information Captured*
Ethnographer	Abstract image of grinders	Provides facts based on observation of the event. Facts include resource type, body mechanics.	Women as food producers Information exchange	Abstract location
Historic Accounts (Spanish)	Abstract image of grinders	Provides material culture facts. Technology is center to body and food production.	Women as food producers Men as entertainment	Household location
Historic Accounts (Crow-wing)	Thinking about the rite of passage while performing related duties	Provides social information: clan membership, relationship between actors. Material culture is implicit, but essential.	Women as food producers: corn Kinship structure. Marriage roles. Male relatives support the event.	Imagined space of grinder Community & Ritual Rite of passage
Corn Narrative	Provides insight into ancestral space (katcina) and symbolic (Spider Woman). Real possibility of death in failure.	Material culture is explicit: ground stone tools and ice.	Solitary grinding, but girl is becoming part of a new community: her in-laws. Social transformation processes for grinder and participants	Imagined space, symbolic Ancestors & Community Ritual Rite of passage

Acorn Narrative	Provides insight into mother's protection of children. Status is associated with skill.	Material culture is implicit.	Role of mother's family in the underworld. Land tenure? Marriage commitment includes level of grinding skill.	Imagined space, symbolic Ancestors & community Ritual & moral implications
Ethnographic Images	Static image of grinders	Image illustrates methods, resources, and material culture.	Women as food producers	Location real & abstract Open air/fixed
Hopi Images	Grinders are Corn Maidens.	Material culture is symbolic and ritual.	Material culture painted with symbols: birds; rain clouds. Literally frames the Corn Maidens.	Imagined space Ancestors & community Ritual space
Archaeologist		Female gendered space Hunter-gatherer or agriculturalist tech Settlement size estimates Environmental adaptation Subsistence pattern	Assumed female space.	Regional patterns Landscape Static

resources or even the technology used for food processing. However, both food and technology get caught up in the performance of the social life history.

Bedrock ground stone features are the workspace of pubescent, adult, and mature individuals. Gender distinctions include all who adopt the grinding tasks, thus inheriting the knowledge and space (see Gayton 1948a, 1948b). Grinders work in groups unless the activity is part of a larger ceremonial context or is a grinding ordeal intended to negotiate the social position or identity of the worker. The framework accepts that a formal kinship relationship exists between the workers at any given space and that ground stone features are culturally related to one another on the landscape. Groups using these sites are still highly mobile, but reliance (socially and culturally) on the bedrock ground stone features and their incorporation into socially significant rites of passage tie them to specific spaces on the landscape.

Regional morphological variation between permanent grinding surfaces may indicate the tool kit and food recipe preference (i.e., mortars for pounding versus metates for grinding) instead of the nature of social grinding processes and the importance of the spaces to social reproduction. The presence of regularly spaced bedrock ground stone features would indicate reliance on critical staple foods, but those foods could be locally collected or grown nearby. Finally, the multi-vocal framework considers the nature of the stories told about these spaces. They indicate a semiotic or symbolically meaningful place for cultures that inhabit these spaces. In addition to organizing social life, experiences, and landscape use, these spaces also place the society into the emotional-spiritual realm (Figure 8.6). Permanent ground stone sites organize more than just landscape procurement strategies. They may, in fact, point to the socialization of the landscape (different spots for ceremonial events, feasts, or quotidian food processing) and the navigation of a world centered at permanent grinding spaces.

SOCIAL REPRODUCTION ON THE CHAQUAQUA PLATEAU

Frame analysis supports the expectation that workspaces, consisting of similar grinding practices, should reflect information exchange between groups of workers (probably women, but incorporates children and other kinsfolk), rites of passages (possibly including secluded areas accessible only to initiates and elders), and because of the ritual nature of some of the spaces, that symbolic representation of ancestral beings or other-world connections would be present. The sites should include spaces for groups of individuals and have multiple ground stone surfaces if they support information exchange. Rites of passage or other ritual contexts are evidenced by repetitious actions. Multiple ground stone surfaces and evidence of reuse coupled with rock art provide evidence, albeit circumstantial, for ritual context and symbolic ideology.

Physical Environment

The BGS features discussed in this chapter are from the JE Canyon survey of a small, spatially bounded, side canyon near the heart of the Chaquaqua Plateau. The canyon can be loosely divided into a north and south surveyed area, with most of the sites along the east-facing rim of the canyon between 400 and 600 feet above the floodplain of Chacuaco Creek. The small tributary canyon lends itself to discussing the social reproduction of culture because of the circumscribed landscape. The eleven bedrock ground stone sites with 207 ground stone surfaces on 23 features (Lynch 2017a) are easily accessible to one another, providing a familiarity of the environment.

Application of Social Reproduction Frames

Results of the frame analysis (Table 8.3) support the assumptions that the bedrock ground stone features were primarily female spaces if they were grinding food resources, but acknowledge that, as part of social ritual, they incorporated the entire community of men, women, and children. Although some of the features served multiple purposes, resource production of the local oak, juniper, and abundant grasses in the side canyon was supplemented by stands of wild fruits (Lynch et al. 2013). Additionally, the spaces were probably occupied by groups of people, including various age categories or social identities (for instance, uncles, aunts, brothers).

Based on the criteria above, the bedrock ground stone features within the canyon were sorted into two groups: those with rock art and those without (Table 8.4). Six of the sites have rock art panels near or superimposed upon the ground stone surfaces. The other five sites are excluded for several reasons, most notably the lack of direct association with the other work areas, such as hearths or campsites. The spatial arrangements of the features and sites in the side canyon could accommodate between two and upwards of 20 people. However, the number of workers possible is constricted by the bedrock features themselves (Figure 8.7). In some cases, only one worker at a time could have ground materials; in others, four or five workers (Lynch 2017a, b). Four of the sites in the canyon have a relatively high concentration of ground stone surfaces (between 24 and 68 surfaces) which could support collaborative work processes. The site

Table 8.4. Summary of Archaeological Sites with Interpretation Based on Frame Analysis

Site	*BGS element (N)*	*Site Area (m^2)*	*Archaeological Features*	*Rock Art*	*Symbols*	*Social Information*	*Spatial context*
5LA11453	68	252	Dry laid walls	1 panel 7 motifs	Abstract linear motifs	Large space to accommodate 4 or more workers; motifs are pecked over the grinding basins	Expansive view of the canyon but sheltered grinding.
5LA11446	29	325	Dry laid stone wall	1 panel 26 motifs	Hunting scene with large & small ungulates. Some tools (atlatl?) superimposed	Rock art panel is part of the site. Grinding feature overlooks the panel. Space accommodates 3-4 grinders.	Expansive view of Chacuaco Creek to the north. Site is very sheltered.
5LA11455	28	1115	Lithic scatter	2 panels 3 motifs	One quadruped, a wavy line and abstract motif	One space is open but could only have two adults, space ideal for children. Second shelter is protected, but low ceiling indicates some grinding could be children.	Both ground stone features overlook the mouth of the side canyon and the Chacuaco. A campsite or tool processing site is on the terrace above this one.

5LA11451	24	270	Tool grooves, hearth, lithics	3 panels 20+ motifs	Motif include multiple ungulates, circular motifs, wavy lines.	There are open spaces and protected ground stone features. Ritual use is suggested by the abstract rock art, small basins, tool grooves, position near water, hearth features.	There are three levels to the site, a deep rockshelter with multiple activity areas, a ledge with multiple ground stone features and rock art and a barren rockshelter with seeps. With a 5 minute walk to spring.
5LA11454	6	88	Two large fixed metates	1 panel, 6 motifs	Rock art motifs are connected curvilinear.	Space is open but could accommodate multiple workers. The rock art panel and fixed metates are sheltered.	Unsheltered ground stone are on a ledge overlooking the Chacuaco. The terrace between the rock art panel and ground stone feature is boulder strewn.
5LA11450	1	200	Debris midden, ground stone fragments	2 panels 38 motifs	Complex curvilinear motifs, circular motifs, human and animal representations	A single ground stone feature next to the main rock art panel.	

areas are large enough to accommodate aggregates of people who could supervise, work together, and exchange information. Although rites of passages were probably conducted in secluded areas, most of the other features on sites suggest multiple concurrent activities.

Ritual use of the space, for rites of passage or first fruits ceremonies, should have a symbolic atmosphere. In the absence of stories, rock art panels serve as a proxy for the symbolic use of space. To follow this line of thought, we have to assume the rock art and ground stone features are contemporaneous. Ritual processes are repetitive, something that could be supported by the number of bedrock ground stone surfaces. Nearby springs at the head of the canyon and the secluded nature of the rocky ford also support the idea that at least this part of the canyon could have been used as part of a ritual system. The presence of rock art at all of the sites with the densest concentration of ground stone supports both the symbolic nature of grinding, at least at some of the sites, and the strong possibility that the ground stone sites were centers of social reproduction.

CONCLUSION

In regions where bedrock ground stone spaces are a part of the archaeological material record, they have the potential to add to our understanding of subsistence patterns, social organization, and landscape use among a host of other cultural processes. Unfortunately, in those without ethnographic examples of the connection of ground stone features with other human systems, the ground stone areas are relegated to the sidelines of archaeological inquiry. The construction of a multivocal framework for application to archaeological materials in these orphaned contexts facilitates insight into how prehistoric peoples inhabited these spaces.

What is often neglected is that the activities associated with ground stone surfaces represent complex social organization, stable resource procurement, symbolic and meaningful social processes, and represent a fundamental attachment to the landscape. In areas without ethnographic evidence, we may not be able to determine the food resource because few residues remain on exposed bedrock surfaces, but the archaeological inquiry has the advantage of looking beyond ethnographic experience to consider material cultural trends from other archaeological contexts (Wobst 1978). The construction and maintenance of social spaces at bedrock ground stone spaces, once properly documented, can answer questions from any region. The approach used here illustrates the importance of ground stone material culture to the broader community and the creation of social networks that transcend subsistence patterns (acorn gathering or corn growing) and food resources (local resources or cultivated ones).

A multivocal representation of bedrock ground stone archaeology seeks to include the participants' experiences and their community (past, present, or future) in the reading and analysis of material culture. Frame analysis provides a method to merge multiple traces of material culture instead of capturing one snapshot of behavior. The model retains positivist methods to explore technological and functional aspects of the bedrock ground stone features and their distribution through time and space but seeks to expand the range of archaeological inquiry to include how space could have been incorporated into the broader community, and social frameworks, particularly in material culture relates to social reproduction processes in the past. The inferences include the experiences found in narratives of lived-in ground stone spaces.

The archaeological implications of a multivocal view of bedrock ground stone features lie in the interpretative framework. The material record—especially in areas that do not have an ethnographic record—lacks the framework to identify with the construction of social space. However, as constructed space, BGS features offer more information about the past using a model of space that supplements techno-functional interpretations of bedrock ground stone features. The model encourages an archaeologist to think of the organization and distribution of BGS features in terms of the society that created and lived in them.

IV

FUTURE RESEARCH

9

Lesson Learned and Future Avenues to Explore in High Plains BGS Features

This book defined two main challenges for the archaeology of bedrock ground stone features. The first is about the nature of archaeological work. In particular, how some aspects of past material culture became invisible and what methods are appropriate in bringing those "ignored" things into the forum. The second explored the dynamic of human cultural behavior, technology, and environment. Developing a critical dialogue about bedrock ground stone features urges us to engage with both of that research simultaneously. However, there remains much work to be done.

THE CHALLENGE

Bedrock ground stone features are not *invisible*. Anyone who has spent time in the canyons of the Chaquaqua Plateau will readily recognize the features, especially those that resemble the Southwestern oval or trough-shaped metates. Most will recognize the similarity between these features and the California acorn mortars integral to social production within many Native American cultures. Others will wonder how the BGS features relate to those found in the eastern United States. Still, the BGS features remain more or less disconnected from the rest of the Colorado archaeological context. The preceding chapters began piecing together the disparate evidence gathered so far into a single intelligible portrait of BGS features within the region. As with all research, more questions have been raised than answered. Some intuitive leaps had to be made, but none less plausible than others made in advancing human knowledge.

The existence of bedrock ground stone features is not in question. Instead, we ask how we should describe and interpret them. A parsimonious path has been taken. We now have enough evidence to begin the journey of discovery. The 462 sites that have been used to develop a description of regional BGS features represent a small fraction of the total possible features. We may never know the actual quantity, but the level of description in field paperwork provides us with an appropriate sample to begin.

CURRENT KNOWLEDGE ABOUT BGS FEATURES

The bedrock ground stone features found on the Chaquaqua Plateau are a unique material culture phenomenon occurring throughout the open landscapes of the local canyon system. The BGS features are also found in the recesses of the myriad side canyons and gulches that drain the plateau. The features are of three main types. The most common, west of the Purgatoire, is the relatively flat, amorphous grinding elements with open views and little shelter. The relatively flat grinding elements found in open-air contexts but near rockshelters range in shapes, sizes, and depths. The least common of all is the widely distributed sheltered BGS features that are found within rockshelters. These spectacular types of BGS feature capture the imagination of everyone who has seen them. They are characterized by multiple grinding surfaces on each outcrop of bedrock. These grinding elements often occur in pairs or clusters, so they were likely used by groups of people to transform raw materials into culturally acceptable products. While the probability is that the raw materials were locally harvested plants, some of the features may have been used by children; some may not have. In contrast, others may have ceremonial possibilities, so quotidian household food production cannot be assumed.

Ethnography and oral tradition from other areas of the United States and grinding cultures throughout the world revealed two landscape strategies employing BGS features: resource processing on the open plains or open areas within the canyons. The people using them would have been in the area for relatively short periods before moving to other parts of the landscape. The BGS features represent family-level social production processes; mother-child, father-mother, perhaps grandmother-grandchild. Those found near rockshelters or within them would have been used for various processing tasks, including seasonal ceremonials or rites of passages. They represent the more considerable group cohesion, where extended families or friend networks gathered to exchange information, gossip, and stories.

The BGS features occurred regularly from the Middle Archaic, about 5,000 years ago, but may have been occasionally used earlier. The height of their popularity appears to have been the Late Archaic to the Late Prehistoric, about 2,000 and 600 years ago. However, during the Apishapa and Sopris periods, the features appear to have been used less frequently. The BGS features seem to have been replaced wholly or partially about 500 to 900 years ago, at least at the periphery of the plateau at sites like Trinchera Cave. If this pattern repeats elsewhere in the region, it could support the hypothesis that local demographic pressures increased as the Pueblo great house cultures declined in the Southwest or as the regional climate became drier.

BGS FEATURES: A LIFE HISTORY

BGS features probably go through many phases, from creation to abandonment. A typical life history sequence might resemble the following. First, an individual recognizes the appropriate type of material useful for the grinding task at hand, perhaps on a boulder, perhaps on an exposed area of bedrock on the open plains. The decision is made to settle down in the near vicinity, in conjunction with other factors, such as water, view, other resources, or socio-cultural factors. Next, raw materials (vegetation, meat, or other resources to be processed) are gathered. Minimal shaping is needed to begin work: the upper rough cortex has to be somewhat cleaned of dirt and debris, but the roughness of the surface is not necessarily an issue. However, this means the sandstone will ground along with food resources. The surface is used until it is no longer needed, either because the camp needs to be moved. Weathering processes take over, causing more surface changes. Open-air features may have been returned to on a seasonal basis. Those in the rockshelters have the same basic sequence. However, ritual use or domestic use could add more layers to the palimpsest.

BGS FEATURES: A SOCIAL HISTORY

The stories told about grinding space incorporate many references helpful in constructing the social history of BGS features. In the Bear and Fawn stories discussed earlier, the literary device portrays the bedrock mortar as a portal to the mother's family in the world of ancestors. The same message can be reversed to an image of the birthing process or a place where social identity is born and nurtured through maternal lineal descent. In another acorn text, At Mīmetaktit She Pounded Acorns (Goddard 1904:185), a mother ignores her hungry child as she works. The child wanders off to the forest, where he is raised by the elements and spirits. His father follows him, always just a village behind until finally, when the boy is almost grown, they are reunited. Returning home, they find the mother has died; with no other family relationship to maintain, they leave to establish a new home elsewhere. Heroic adventures aside, the story provides many levels of social identities. Both social and physical death results from activities grinding space.

The richest social history lessons associated with permanent ground stone spaces move a grinding person from one stage of social identity to another. These ceremonial grinding activities are attached to puberty ceremonies (biological changes that necessitate a cultural bridge) and marriage ceremonies (creating new family identity and new lineages). While peripheral to most of these ceremonies, men generally assume supportive roles by providing for feasting events. In many grinding ceremonials, only the best grinders are allowed to enter this spiritual realm.

Children are an integral part of this space. Woodbury (1954) indicates some of the bedrock grinding elements he observed were credited to children. Whether watching, pestering, or playing, children learn their place in the social and physical world from their proximity to grinding activities. The expectation for finding adolescent or pre-pubescent play spaces within the spectrum of BGS features is only now becoming part of the quest to describe and understand these types of grinding workspaces. If they are only more miniature versions of the same adult shapes, we can use the morphometric analysis from chapter 4 to look for the presence of children through the size and shapes of the grinding elements. If children were nearby, this might account for some size variation between BGS elements on the same feature. Like today, children often practice their skills away from the watchful eyes of adults, a point made by Woodbury (1954).

As with most social activities, food processing is rarely static. The quotidian activity provides space for people to gather. In this space, identities are traded along with news and local gossip. Friendships are formed, nurtured, and maintained—even the most ordinary food preparation connected people to their ancestors and gods. The value of an individual's labor is converted into a tangible commodity, but as with all daily repetitive action, there is a fundamental meditative state that accompanies it. Within this socio-cognitive state, people become more attached to the technology that enables them to make and maintain these connections and thus become more resilient.

Resilience at the most superficial level is the ability to recover from setbacks or unplanned events quickly. These could be environmental fluctuations such as the loss of once stable food resources. The ability to shift with changing ecosystems is supported by the flexibility of social life and group identity. The environment may change, but the group dynamic adjusts because the social resilience united with technological capacity and intimate landscape knowledge provides a cognitive cushion.

The dynamic system of technology, practicality, and the development of social personhood supports behavioral resilience. The regional peoples valued the bedrock ground stone features over other types of portable ground stone tools. Some sites may have been more practical, while others were symbolic. However, each type of BGS features played some part in the social lives of the people who used them. Ultimately the technology served a social purpose that supported long-term resilience, changing only as the social needs required or allowed. The entanglement of material things (*sensu* Malinowski 1922) with past social lives is difficult to illustrate, except through inclusive methodologies, most of which are still evolving. However, 3D visualization and virtual reality provide the means to do support the process.

QUESTIONS ABOUT THE LANDSCAPE CONTEXT

The evidence indicates that BGS features are primarily found in two primary landscape contexts. The first is open-air. These grinding elements exhibit little shaping or depth, so they were probably lightly used, perhaps only for a season or two. Alternatively, the features may have been used for only a few hours at a time but over many years. The second type of BGS feature is found in or near rockshelters. The grinding elements were intensively used for more prolonged periods of time. However, the shape variation is more extensive when grinding elements are within a rockshelter. The features near but outside of rockshelters are characterized by little shaping or depth, resembling those in other open-air contexts. What is responsible for these patterns? Do the differences between the landscape contexts represent different occupation episodes, illustrating an evolutionary sequence of human behavior? How do we explain the relatively consistent pattern between the Archaic and Late Prehistoric, in which there appears to be increased use of open-air features?

At present, answers to these questions would be speculative. While we have images of open-air BGS features, there are few reliable metric data sources to analyze them. Employment of the variables presented in earlier chapters to analyze the morphology of these shapes requires additional planning. Of the five shape variables discussed earlier, perhaps only those characterizing the size of the grinding elements would apply to answering how BGS elements differ in the landscape contexts. Newer methods of shape analysis must be developed to account for the amorphous characteristics of open-air BGS features. The best methods will incorporate 3D models that can be widely tested using analyses morphometrics (Lynch forthcoming).

The data for the open-air features from the JE Canyon sample offers some insight into the difference of grinding elements between landscape contexts. Although the open-air grinding elements are near the BGS features at 5LA11446 (a rockshelter with many grinding elements), we cannot date them. The open-air BGS features appear to differ significantly from those found in the PCMS sample. The grinding elements from the JE Canyon sample have oval shapes with depth, while those on the PCMS are amorphous with little or no depth. Only one site in the JE Canyon sample (5LA11454) has amorphous bedrock ground stone elements, eclipsed by the presence of oval-shaped elements on the same BGS feature. Photogrammetry or laser scanning of the open-air BGS features would be the best means to expand our understanding of their morphology and how they fit into the broader landscape, as well as their role in cultural change on the Chaquaqua Plateau.

Additional questions arose about the relationship between site contexts and tool types. In particular, whether the relationship indicates a long-term relatively stable adaptation to the local canyon environment. Otherwise, it reflects a hedging-our-bets strategy by adopting a range of tool types. To explore adaptive and evolutionary processes, neighboring site types have to be considered part of a landscape system that included the bedrock ground stone features.

Without direct dating of the BGS features, we can only generalize about their use through time. For example, Archaic sites tend to be larger than those dating from the Late Prehistoric. However, many of the Archaic sites continued to be used during the Late Prehistoric, suggesting that as landscape knowledge became more intimate, the sites expanded accordingly. The familiarity and intimacy of the landscape may have fostered the development of symbolic use of some sites with quotidian use of others. These questions require a better understanding of the individual BGS feature elements, thus underscoring the need for consistent site recording processes.

We can offer some discussion about how the different landscape contexts might represent two types of land use: one that is seasonal and quotidian, the other seasonal but ceremonial. The large open-air sites and those near rockshelters might represent the ordinary use of the on-site resources. At the same time, people go about their business each season, gathering and processing local raw materials. There appears to be little difference in the relative use of open-air or sheltered sites, based on the elevation of sites, but prehistoric sites have a greater range of elevations. Open-air sites are generally further from modern permanent water sources. Archaic sites, on average, are closer to water than those occupied during the Prehistoric. The critical factor seems to be the intensity (represented by the number of grinding elements and their shapes) of grinding within rockshelters instead of open-air contexts.

ARCHAEOLOGICAL CONTEXTS

The archaeological contexts provide more information about the other features found with grinding elements. The two main types of contexts are camps or stone structures. The latter is too broad a category to develop inferences since the structures are circular or linear stone alignments. In rare cases, the stone structures are contiguous vertical slab features typical of the Apishapa culture. If we look, as an example, at the intensive survey data from JE Canyon, then the stone structures found near and on sites can be interpreted as foundations for storage facilities or small housing structures. These rectangular-shaped features are found at the head of the surveyed side-canyon as well as the canyon rim overlooking Chacuaco Creek (Lynch, forthcoming). On the other hand, greater stone structures were located adjacent to the BGS features sites with higher grinding elements and rock art panels. Why? Is there a bias in the recording process, or is there a spatial pattern that needs to be further explored?

An example of recording bias is in the JE Canyon survey. It was focused on describing the rich quantity of BGS features sites, not the other archaeological features. For instance, we recorded diagnostic artifacts, an Archaic projectile point found at the head of the canyon with other lithic materials and stone alignments. We did not collect or excavate at any sites, including non-BGS feature sites, although we did seek to tie the sites together into a coherent story of land use. Thus, while we have a general idea of the range of archaeological features, we can only guess about the relationships between the BGS features and other non-BGS feature sites in the same side canyon.

Additionally, there are probably features within the Chacuaco floodplain related to the workspaces found on the upper terraces of this side canyon. However, the survey data we have resulted from a pedestrian survey that only followed the sloping landscape for the upper canyon rim. Thus, if vertical landscape use was in play, additional surveys must be completed in the floodplain and the neighboring canyons to describe it.

Another type of archaeological feature often found in association with the BGS feature is rock art. These sites are located mainly in sheltered contexts, although some open-air features also have rock art (5LA10934). However, rock art is present at only 23 percent of all the sites with BGS features. Many are sizeable complex rock art panels like those at JE

Canyon (5LA11450, 5LA11454, and 5LA11451), Metate Cave, and Trinchera Cave. Conversely, small panels are composed of single or repetitious symbols in the JE Canyon sample (5LA11455, 5LA11453). If we use rock art to link socio-cultural symbolic behavior with more ordinary routines such as grinding, we can infer that these patterns suggest some sites are more ceremonial while others are not.

OTHER POSSIBILITIES

This book has taken a fairly conservative approach to consider the nature of BGS features. Nevertheless, there are examples of other possibilities that would complicate the interpretation of these features. We will consider two here: children and song. Both are part of the social nature of grinding spaces.

Children often "play" on nearby rocks while the adults work at the BGS feature. What would children's workspaces look like? Would adolescent work-play leave behind smaller grinding elements or play grinding elements (Woodbury 1954) and abstract or non-intentional rock art motifs? Other types of cultural practices not related to resource processing could leave behind grinding-like elements associated with rock art. An example of these includes the "rock gongs" in Africa (Goodwin 1957; Jackson et al. 1965; Jellicoe 1965).

Soundscapes are important to grinding work, especially to intensive or ceremonial grinding. Creating various types of music might also leave behind trace elements in the local stone. The closest ethnohistoric evidence we have for sound, in the form of music, being associated with grinding is from descendent communities of Athapaskan-language speakers who may have moved through or inhabited the Chaquaqua Plateau at some point around the time the BGS features were abandoned (Seymour 2012). As a relatively mobile presence on the Puebloan frontier, the Athapaskan speakers would have been, and still are, an essential part of the landscape. Unfortunately, the sparse records render the social grinding space of the Jicarilla, one of the descendant communities, the most difficult to reconstruct. Opler (1936: 324) describes corn grinding songs to accompany grinding corn. "The Jicarilla possessed corn-grinding songs and made something of a social occasion of the work" (Opler 1936:324). These songs were designed to be sung to the group of women doing grinding by a group of older men of the community.

Grinding is not the solitary work of a nuclear family or single woman. Instead, it falls into the domain of the extended family and community-at-large, much like other Puebloan groups. The words of the songs seem designed to tease the workers, "'Hurry up, hurry up, grind. Your husband is hungry!'" (Opler 1936:324). Opler does not tell us if these songs are serious, but one gets the impression from his description that the mood is joking and lighthearted. The girls are being teased. Neither Russell (1898) nor Goddard (1911) provides songs for social grinding, although one consultant, Casa Maria, provided a corn mush and corn soup recipe complete with grinding instructions (Goddard 1911: 262–263). It may be that these songs had become so obsolete by the time they were recorded that only a few people remembered them, and then gradually, through disuse and time, they have become a ghost of the grinding space (Kurath, 1960:6).

For example, grinding spaces are more than sets of technological advances that allowed for large centers of corn flour processing. They are spaces that begin and end in the imaginations and memories of the peoples who are integrated into the transformative act of grinding. The Jicarilla, displaced from their eastern territories, still retain through their grinding songs and origin stories a grinding space. Though their processes have changed and may appear disconnected from their landscape, through the remaining songs and stories, Jicarilla people retain access to those grinders in their past much the same as modern Hopis retain a link to their distant relatives through the corn Kachinas.

Although grinding spaces have been ritualized in the southwestern United States, this is not to say that all cultures ritualize their grinding spaces. For example, the Hadza people who use exposed bedrock to grind food do not have work songs or singing processes associated with their work (James O'Connell personal communication 2020). Likewise, this is not to say that social reproduction processes are absent or somehow less rich than other grinding cultures. It is only to point out that differences may shed some light on the two seemingly disparate examples between open-air and sheltered spaces.

On the one hand, we have open-air spaces with little evidence to help understand how they fit into regional archaeology. On the other hand, only 50 percent of the entire sample of BGS features sites have any material remains other than the bedrock ground stone feature. Why? Perhaps other types of material culture were not reported, or they may be absent altogether. If the latter scenario is true, then BGS features might have been created just far enough from other activities, like a camp, that they were recorded as a different site. In that scenario, the other types of activities associated with grinding would not be included in the data.

Further research would include broadening the sites sampled to those further than 50 meters and 200 meters away from the BGS feature site. Using knowledge of the spatial distribution of Hadza sites as an example, a new GIS study is underway to do just that. In addition, the new study investigates the distribution of other areas found in the neighborhood of the BGS features, using the grinding spaces as an anchor for other activity areas. The challenge will be placing the features in time to investigate changes in landscape use patterns.

NATURAL AND HUMAN AGENCY

One challenge in studying BGS features without the corresponding ethnographic context is determining human agency from natural processes. McCarthy and colleagues (1985) reported a situation in which Native consultants dismissed elements archaeologists recorded as natural, not cultural. For example, weathering elements such as freeze-thaw or wind-water erosion create many shapes on bedrock outcrops. Naturally shaped features can be squarish, oval, circular-shaped, or amorphous (Lynch forthcoming). So how do we verify our field identification of BGS features?

A closer look at the reports and images for BGS feature sites provides clues about how archaeologists make the determination, in the field, between human and natural processes. Although some field notes discuss the location of the ground stone elements as being in areas that are subject to wind-rain and freeze-thaw processes, most go by the feel and look of the surface. Hunters and gatherers who use bedrock outcropping rocks to process food resources leave a specific type of pattern behind (James O'Connell, personal communication 2020). The patterns are an elongated surface about 30–40 cm across. Food residues are often left behind, but most of the materials are scraped clean for consumption. While there is little surface preparation other than pounding or pecking the surface to create a roughened area that helps break down the food. The resulting pattern is recognizable as similar to other grinding stones, the most notable being from the southwestern United States.

Consultants from descendant Native American communities have pointed out the similarities of the BGS features to places where women ground things (Stoffle 1994). Ethnohistoric recognition of these types of grinding spaces is the closest we will get to ethnographic context. Examples of the latter would be Gayton's (1948 a, b) work in California. On the other hand, ethnography documents what is remembered or the stories that are told about events, material culture, or landscapes.

Use-wear studies on ground stone tools are most helpful in determining the difference between natural processes and human agency (Adams 1993, Owens 2006). These studies illustrate that shape alone is not the best determinant for functional assessment but instead looks closely at the polished area's nature for specific patterns of the sandstone matrix. While best done using a microscope in a laboratory, some patterns are discernable in the field. More sophisticated microscopy is becoming available, and we will soon be able to test the wear assessments on BGS features. We have yet to determine if 3D models can be used to look at wear patterns. Finally, although Adams (1993) and Owens (2006) successfully conducted lab analysis to determine use-wear patterns, there are no local field experimentations specific to BGS features. True (1993: Fig. 3:6) shows slicks or areas of the bedrock surface that have been smoothed by grinding or moving resources across the stone surface. Although speculative, the open-air features might be winnowing areas or pre-processing areas. If so, the open-air features might be very short-term occupations. These types of sites may not be associated with campsites or other activities areas. The closest ethnographic context for flat grinding elements is from southern California, where Kroeber (1909:51) writes, "a metate is nothing but a flat stone, oval or somewhat rectangular in shape." Perhaps then, any flat surface with ground use-wear should be considered a food-producing or resource-processing area.

We propose several avenues of research to understand the open-air BGS feature. The first is a long-term observational study about signatures left by natural processes. Additionally, more ethnoarchaeological field studies with extant grinders such as the Hadza would provide modern analogies for morphometric comparison such as that in chapter 2. Finally, we should not exclude the possibility that some of these open-air features may have been used to process other resources. An example of this could be that they were used to process (beat or pound) yucca leaves to break down the fibers for creating rope-like materials (see Reynan and Morse 2016:99). Ground stone tools are resilient technology in that they are multi-functional, thus, highly adaptive. This type of problem requires an interdisciplinary and multivocal approach to material culture.

PLOTTING A COURSE FOR BGS FEATURE ARCHAEOLOGY

The inclusion of BGS features into the broader regional archaeology enriches our knowledge of the ancient peoples of the Chaquaqua Plateau and the neighboring Rocky Mountain – Plains archaeology. This book has provided evidence that the features are more than just food-processing equipment. They can be part of embedded socio-cultural production processes. Far from being enigmatic, there is much to be added to our knowledge of the past from their study and inclusion in broader archaeological discourse. However, BGS feature research is still in the nascent stages. Many of the hypotheses within this book have yet to be rigorously tested.

Perhaps the most significant gap in our knowledge is the lack of direct connections between modern archaeology and the descendant communities whose ancestors inhabited the region. While the Jicarilla or the

Southern Ute were probably the people who lived in the region during or very near to the end of the abandonment processes, there is little information on how the BGS features would have been used. The Navajo, another member of the Athapaskan language family, may have been the region. Perhaps the closest living relatives to those who created the BGS features would be the Arikara-Pawnee (Echo-Hawk 2000). We will know precious little until long-term ethnohistorical projects are completed (Lynch forthcoming).

In the meantime, several research projects are being carried out to propel Colorado BGS features into the next critical research phase. While we have begun the process of placing the features within the context of other neighboring features, artifacts, and landscape contexts, there is still so much to learn. A new project is underway that uses Geographic Information Systems (GIS) to model BGS feature sites with their nearest neighbors. This research aims to compare material culture patterns of modern hunter-gatherers who use bedrock to process resources with those found in the Chaquaqua Plateau. The usefulness of the "site" as a heuristic device for understanding BGS features may be obscuring their presence by enforcing exclusionary practices. Thus, they are more easily ignored if they are not explicitly queried, as done in previous chapters. The new project will build frames of reference that include multiple site categories.

The research presented in chapters 4, 5, and 6 is the current results of 3D visualization research at the time of writing. New research has already begun to develop metric analysis procedures for measuring 3D modeled grinding elements. The goals of the new research are two-fold. The first is to build a more comprehensive method for analyzing the shapes. Only three axes are used to understand, abstractly, the differences between shapes and sizes. However, current research uses four planar axes and five depth axes to develop new shape models that will refine our understanding of shapes and their distribution in the region. The second provides digital data sets for broader collaboration between scholars engaged in bedrock ground stone research.

A new project is underway to develop methods that measure use-wear on the grinding elements. It is imperative to refine our knowledge of the open-air BGS features to determine the erosional processes that operate on them post-abandonment and understand the differences between natural and human agency. While those BGS features exhibiting typical shapes (ovals, circles, or rectangular) within rockshelters are easily recognizable, those in open-air may be easily skipped over in field surveys. Thus, having more efficient ways to make field identification will help clean up our current database while also allowing more people to "see" the BGS features when they come across them.

Finally, building a more inclusive understanding of the nature of BGS features entails the experiences of Euro-American expansion in the region. Many sites, for instance, Metate Cave, were visited by local ranchers, farmhands, or day-tourists. The names they left behind can build a new narrative about the ordinary lives of ranchers, farmers, and even schoolteachers. These experiences are part of our shared cultural heritage. One of the most important lessons learned from studying BGS features is that they are rarely used to exclude people from the social reproduction of knowledge. On the contrary, they are hard taskmasters against whose stone humans ritually grind themselves to reinvent or renew their social identities. These ancient spaces belong in the archaeological mainstream, along with their resistant, stubborn natures. Bedrock ground stone features are places that connect people to the landscape, and through shared experiences, to one another and our ancestors.

References

Adams, J. L., 1993. Mechanisms of Wear on Ground Stone Surfaces. *Pacific Coast Archaeological Society Quarterly* 29(4): 61–74.

Adams, J. L., 1999. Refocusing the Role of Food-Grinding Tools as Correlates for Subsistence Strategies in the U.S. Southwest. *American Antiquity* 64(3): 475–498.

Adams, J. L., 2002. *Ground Stone Analysis: A Technological Approach*. University of Utah Press, Salt Lake.

Ardener, S., 1981. Ground Rules and Social Maps for Women: An Introduction. In *Women and Space: Ground Rules and Social Maps*, edited by S. Ardener. New York: St Martin's Press, 11–34.

Asch, M. I., 1973. An Anthology of North American Indian and Eskimo Music. In *Folkways Records Album No. FE 4541*, edited by Michael I. Asch. Folkways Records and Service Corps, New York.

Bakhtin, M., 1981. *The Dialogic Imagination: Four Essays* (University of Texas Press Slavic Series). Austin: University of Texas Press.

Barrett, S. A., and Gifford, Edward W., 1933. Miwok Material Culture. Bulletin of the Public Museum of the City of Milwaukee 2(4): 127–277.

Bartlett, K., 1933. Pueblo Milling Stones of the Flagstaff Region and their Relation to Others in the Southwest: A Study in Progressive Efficiency. Museum of Northern Arizona Bulletin. Northern Arizona Society of Science and Art, Flagstaff.

Basso, K. H., 1988. Speaking with Names: Language and Landscape Among the Western Apache. *Cultural Anthropology* 3(2): 99–130.

Basso, K. H., 1996. *Wisdom sits in places: Landscape and language among the Western Apache*. UNM Press.

Beaglehole, E. and Beaglehole P., 1937. *Notes on Hopi Economic Life*. (Yale University Publications in Anthropology No. 15.) New Haven: Yale University Press.

Bennyhoff, J. A., 1953. High Altitude Occupation in the Yosemite Park Region. Some Archaeological Sites and Cultures of the Central Sierra Nevada: 31–32.
Bennyhoff, J. A., 1956. An Appraisal of the Archaeological Resources of Yosemite National Park. Reports of the University of California Archaeological Survey. Berkeley.
Binford, Lewis R., 1980. Willow Smoke and Dogs' Tails: Hunter-Gatherer Settlement Systems and Archaeological Site Formation. *American Antiquity* 46(1): 4–20.
Black, K. D., 2000. Lithic Sources in the Rocky Mountains of Colorado. *Intermountain Archaeology*, pp. 132–147.
Bourdieu, P., 1990. *In Other Words: Essays towards a Reflexive Sociology*. Stanford: Stanford University Press.
Bourdieu, P., 1991. *Language and Symbolic Power*. Cambridge: Harvard University Press.
Bourdieu, P., 1998. *Practical Reason: On the Theory of Action*. Stanford: Stanford University Press.
Buonasera, T. Y., 2012. Expanding archaeological approaches to ground stone: Modeling manufacturing costs, analyzing absorbed organic residues, and exploring social dimensions of milling tools.
Buonasera, T., 2016. Lipid residues preserved in sheltered bedrock features at Gila Cliff dwellings National Monument, New Mexico. J. Lithic Stud. 3(3), 78–101.
Campbell, R., 1969. Prehistoric Panhandle Culture on the Chaquaqua Plateau, Southeast Colorado, Department of Anthropology, University of Colorado.
Campbell, R., 1976. *The Panhandle Aspect of the Chaquaqua Plateau*. Lubbock: Texas Tech Press.
Cassells, E. S., 1997. *The Archaeology of Colorado*. Johnson Books.
Claassen, C., 2011. Rockshelters as Women's Retreats: Understanding Newt Kash. *American Antiquity* 76(4): 628–641.
Claassen, C., 2016. Native American Landscapes: An Engendered Perspective. The University of Tennessee Press, Knoxville.
Crow-wing, and Parsons, E. W. C., 1925. *A Pueblo Indian Journal, 1920–1921*. (Memoirs of the American Anthropological Association.) Menasha: American Anthropological Association.
Dangel, R., 1929. Bear and Fawns. *The Journal of American Folklore*, 42(163): 307–308.
David, N., 1998. The Ethnoarchaeology and Field Archaeology of Grinding at Sukur, Adamawa State, Nigeria. *African Archaeological Review* 15(1): 13–63.
Densmore, F. 1915. The Study of Indian Music. *The Musical Quarterly* 1(2): 187–197.
Dick, H. W., 1946. Archaeology of the Upper Rio Grande Basin in Southern Colorado and Northern New Mexico. E. B. Renaud. (University of Denver, Department of Anthropology, Archaeological Series, Sixth Paper, 44 pp., Denver, 1946.). *American Antiquity*, *12*(2): 130–131.
Dick-Bissonette, L. E., 1998. Gender and Authority among the Yokuch, Mono, and Miwok of Central California. *Journal of Anthropological Research*, 54(1): 49–72.
Dill, C. E., 1975. Mortars and Their Measurements. In *Papers on the Chumash*, pp. 91–113. San Luis Obispo County Archaeological Society Occasional Paper Number Nine.
Driver, H. E., 1953. The Acorn in North American Indian Diet. Proceedings of the Indiana Academy of Science 62: 56–62.

Driver, H. E., and Massey, W. C., 1957. Comparative studies of North American Indians. *Transactions of the American Philosophical Society*, New Series 47(2): 165–456.

Du Bois, C., and Demetracopoulou, D., 1931. Wintu myths. *University of California Publications in American Archaeology and Ethnology*, 28(2): 279–403.

Dubreuil, L., and Savage, D., 2014. Ground Stones: A Synthesis of the Use-Wear Approach. *Journal of Archaeological Science* 48: 139–153.

Echo-Hawk, R. C., 2000. Ancient history in the New World: Integrating Oral Traditions and the Archaeological Record in Deep Time. *American Antiquity*, pp. 267–290.

Eddy, F. W., 1964. *Metates and Manos: The Basic Corn Grinding Tools of the Southwest*. Santa Fe: Museum of New Mexico Press.

Eighmy, J. L., 1984. A Cultural Resource Inventory of the John Martin Reservoir, Colorado.

Fenenga, F., 1952. The Archaeology of the Slick Rock Village, Tulare County, California. *American Antiquity* 17: 339–347.

Fenneman, N. M. 1931. *Physiography of the Western United States*. McGraw-Hill, New York.

Fewkes, J., 1903. *Hopi Katcinas Drawn by Native Artists*. (Bureau of American Ethnology Annual Report.) Washington, D.C.: Smithsonian Institution.

Feyhl, K. J., 1980. Tool Grooves: A Challenge. *Archaeology in Montana* 21(1): 1–32.

Flint, R. and Flint, S. C., 2005. *Documents of the Coronado Expedition, 1539–1542: "They were not Familiar with his Majesty, nor did They wish to be his Subjects."* Southern Methodist University Press, Dallas, Texas.

Franklin, Jay, Lucinda Langston, and Meagan Dennison. 2016. Bedrock Mortar Hole Sites as Artifacts of Women's Taskscapes: Late Archaic and Early Woodland *Chaine Operatoire* on the Upper Cumberland Plateau of Tennessee. In Native American Landscapes: An Engendered perspective, edited by Cheryl Claassen, pp. 38–84. The University of Tennessee Press, Knoxville.

Frisbie, C. J., 1967. *Kinaalda: A Study of the Navaho Girl's Puberty Ceremony*. Middletown: Wesleyan University Press.

Galloway, P. K. 2006. *Practicing Ethnohistory: Mining Archives, Hearing Testimony, Constructing Narrative*. University of Nebraska Press.

Gardi, R., and B. Gardi, 1995. *Momente Des Alltags: Fotodokumente Aus Nordkamerun, 1950–1985 (Tschadsee, Mandara, Alantika)* Museum für Völkerkunde und Schweizerisches Museum für Volkskunde Basel, Basel.

Gayton, A. H., 1948a. Yokuts and Western Mono Ethnography: Tulare Lake, Southern Valley, and Central Foothill Yokuts. *Anthropological Records*, 10(1): 1–140.

Gayton, A. H., 1948b. Yokuts and Western Mono Ethnography: Northern Foothill Yokuts and Western Mono. *Anthropological Records*, 10(2): 143–301.

Gebhard, P. H., 1943. The Excavation of an Archaeological Site on the Purgatoire River, Southeastern, Colorado. *Papers of the Excavators Club*, Vol. 2, No. 2. Excavators Club, Cambridge.

Gifford, E. W., 1916. Miwok Moieties. *University of California Publications in American Archaeology and Ethnology*, 139–194.

Gifford, E. W., 1917. Miwok Myths. *University of California Publications in American Archaeology and Ethnology*, 12: 283–338.

Gifford, E. W., 1923. Western Mono Myths. *The Journal of American Folklore*, 36(142), 301–367.

Gifford, E. W., 1932. The Northfork Mono. *University of California Publications in American Archaeology and Ethnology*: 15–65.

Gifford, E. W., 1936. Californian Balanophagy. In *Essays in Anthropology Presented to A. L. Kroeber in Celebration of his Sixtieth Birthday*, edited by R. H. Lowie. Berkeley: University of California Press, 87–98.

Gifford, E. W., 1967. *The Use of Acorns in California. North American Indians, a Sourcebook*, 403–409. Macmillan, New York.

Goddard, P. E., 1904. Hupa Texts. *University of California Publications in American Archaeology and Ethnology*, Vol. 1, No. 2.

Goddard, P. E., 1906. Lassik Tales. *The Journal of American Folklore*, 19(73): 133–140.

Goddard, P. E., 1911. *Jicarilla Apache Texts* III. Anthropological Papers of the American Museum of Natural History New York.

Godelier, Maurice, 1986. *The Mental and the Material: Thought, Economy, and Society*. London, Verso Publishers.

Goffman, E., 1974. *Frame Analysis: An Essay on the Organization of Experience*. New York: Harper and Row.

Goodwin, A. J. H., 1957. Rock Gongs, Chutes, Paintings, and Fertility. *The South African Archaeological Bulletin* 12(45): 37–40.

Green, J., 1990. *Cushing at Zuni: The Correspondence and Journals of Frank Hamilton Cushing*. Albuquerque: University of New Mexico Press.

Gunnerson, J. H. 1989. *Apishapa Canyon Archaeology Excavations at the Cramer, Snake Blakeslee, and Nearby Sites*. Reprints in Anthropology 41. J & L Reprint Company, Lincoln.

Hammond, G. P., and Rey, A., 1966. *The Rediscovery of New Mexico, 1580–1594: The Explorations of Chamuscado, Espejo, Castao de Sosa, Morlete and Leyva De Bonilla and Humana*. Albuquerque: University of New Mexico Press.

Hammond, G. P. and A. Rey, 1940. *Narratives of the Coronado Expedition, 1540–1542*. Coronado Cuarto Centennial Publications, 1540–1940. University of New Mexico Press, Albuquerque.

Haney, J. W., 1992. Acorn Exploitation in the Eastern Sierra Nevada. *Journal of California and Great Basin Anthropology*, 14(1): 94–109.

Heizer, R. F., 1978. Mythology: Regional Patterns and History of Research. In *California*, edited by R. F. Heizer. (Handbook of North American Indians, Vol. 8, gen. ed. W. C. Sturtevant.) Washington, D.C: Smithsonian Institution Press, 654–657.

Hodder, Ian, 2012. *Entangled: An Archaeology of the Relationships Between Humans and Things*. Wiley-Blackwell, Malden, MA.

Huscher, B. H., 1943. Indian Stone Enclosures of Colorado and New Mexico. E. B. Renaud. (Anthropological Papers, No. 2, University of Denver, Jan. 1942.). *American Antiquity*, 8(4): 404–405.

Jackson, G., Gartlan, J. S., and Posnansky, M., 1965. Rock Gongs and Associated Rock Paintings on Lolui Island, Lake Victoria, Uganda: A Preliminary Note. *Man* 65: 38–40.

Jackson, T. L., 1991. Pounding Acorn: Women's Production as Social and Economic Focus. In *Engendering Archaeology: Women and Prehistory*, edited by J. M. Gero and M. W. Conkey. Oxford: Blackwell, 301–325).
Jellicoe, M. R., 1965. Rock Gongs in Tanzania. *Man* 65: 85.
Johnson, C. I., 1964. Navajo Corn Grinding Songs. *Ethnomusicology* 8(2): 101–120.
Kingsbury, L. A., and Nowak, M., 1980. Archaeological Investigations on Carrizo Ranch, 1974–1979. Colorado College Publications in Archaeology. Colorado College, Dept. of Anthropology, Colorado Springs.
Kroeber, A. L., 1909. Sinkyone Tales. *Journal of American Folklore,* 32: 346–51.
Kroeber, A. L., 1976. *Handbook of the Indians of California.* Dover Publications, New York, NY. Originally published 1925 Bulletin 78 of Bureau of American Ethnology, Smithsonian Institution, Washington, D.C.
Kurath, G. P. 1960. Music and the Daily Bread. *Ethnomusicology* 4(1): 1–9.
Lefebvre, Henri, 1991. *The Production of Space.* New York: Blackwell Publishers.
Leftwich, B. M., 2010. Set in Stone: Bedrock Mortars and Behavior in the North-Central Sierra Nevada. Ph.D. dissertation, Department of Anthropology, University of California, Santa Barbara.
Lévi-Strauss, C., 1969. *The Raw and the Cooked*. New York: Harper and Row.
Loendorf, L. L. 2008. Hunting, Grinding, and Dancing: Petroglyph Sites on the J.E. Canyon Ranch. Colorado Historical Society Colorado Rock Art Association, Denver.
Loendorf, L. L., and Loendorf, C. R., 1999. Archaeological Sites in Welsh Canyon, Las Animas County, Colorado. Midwest Archaeological Center National Park Service, Lincoln.
Lynch, E. M. 2014. Prehistoric Grinding Landscapes of the Southern Plains: The Archaeology of Bedrock Ground Stone Features on the Chaquaqua Plateau, Southeastern Colorado. Ph.D. dissertation, Department of Anthropology, University of Wyoming, Laramie.
Lynch, E. M., 2017a. Bedrock Ground Stone Features on Chacuaco Creek, Southeastern Colorado. *Plains Anthropologist* 62(243): 219–246.
Lynch, E. M., 2017b. Ancient Shapes, Modern Measures: a Quantitative Method to Describe Bedrock Ground Stone Shape. *Journal of Archaeological Science Reports* 13: 211–221.
Lynch, E. M., Holthus, L., Sanchez, K., 2013. Southern Plains grinding landscapes: Can a local wood rat midden inform us about prehistoric human grinding workspaces and behavior at a Colorado rockshelter? Poster presented to the Wyoming Archaeological Society, Annual Meeting, Laramie, Wyoming.
Lynch, E. M., Matthews, N., and Noble, T., 2017. Unravelling the enigma of prehistoric bedrock ground stone features on the Chaquaqua Plateau, using close-range photogrammetry. *Quaternary International* 439(Part B): 50–68.
McCarthy, H., Hicks, R. A., and Blount, C. M., 1985. A functional analysis of bedrock mortars: Western Mono food processing in the southern Sierra Nevada as Appendix F. In *Cultural Resources of the Crane Valley Hydroelectric Project, Madera County, California*. San Francisco: Pacific Gas and Electric Company.
Mohr, A., 1954. The Deep-basined Metate of the Southern California Coast. *American Antiquity* 19: 394–396.

Nowak, Michael, and Heather J. Gerhart, 2001. Archaeological Investigations in Southeastern Colorado. Colorado College Publications in Archaeology, No. 21. The Colorado College, Colorado Springs, Colorado.

Nowak, Michael, and Heather J. Gerhart, 2002. Archaeological Investigations in Southeastern Colorado. Colorado College Publications in Archaeology, No. 22. The Colorado College, Colorado Springs, Colorado.

O'Connell, James F., Hawkes, Kristen, and Blurton-Jones, Nicholas, 1991. Distribution of Refuse-Producing Activities at Hadza Residential Base Camp: Implications for Analysis of Archaeological Site Structure. In *The Interpretation of Archaeological Spatial Patterning*, edited by Ellen Kroll and T. Douglas Price: 58–76. Plenum Press, New York.

Owens, M., 2006. Edge-Ground Cobble Tools of the Pinon Canyon Maneuver Site, Southeast Colorado. *Southwestern Lore*, *72*(3): 1–27.

Owens, M., Loendorf, L., Hill, E., Hughes, R. E., Krause, R. A., and Owens, P. R., 2007. Archaeological Sites Inventory of the High Priority Portions of Training Areas 1, 2, 3, 4, 5, 6, 11, 13, and H of the Pinon Canyon Maneuver Site, Las Animas County, Colorado, p. 268. Ft. Belvoir Defense Technical Information Center.

Owens, M., C. R. Loendorf, V. W. Schiavitti, and L. L. Loendorf, 2000. Archaeological Sites Inventory in the Black Hills of the Pinon Canyon Maneuver Site, Las Animas County Colorado. Midwest Archaeological Center, National Park Service.

Owens, M., and L. L. Loendorf. 2004. Archaeological Sites Inventory of the Training Area 10 and 12 Portions of the Pinon Canyon Maneuver Site, Las Animas County, Colorado. Midwest Archaeological Center, National Park Service.

Meltzer, D. J., 1983. The antiquity of man and the development of American archaeology. In *Advances in archaeological method and theory* (pp. 1–51). Academic Press.

Michlovic, M., 1986. Cultural Evolutionism and Plains Archaeology. *Plains Anthropol.* 31: 207–218.

Noble, T. and N. Matthews, 2016. Expert Workflow. TN Photogrammetry, LLC.

Noble, T. and N. Matthews, 2018. Expert Workflow. TN Photogrammetry, LLC.

Opler, M. E. 1936. A Summary of Jicarilla Apache Culture. *American Anthropologist* 38: 202–223.

Parkman, E. B., 1994. The Bedrock Milling Station. In *The Ohlone Past and Present: Native Americans of the San Francisco Bay region*. Edited by L. J. Bean. Menlo Park: Ballena Press 43–63.

Parsons, E. W. C., 1930. Zuni Tales. *The Journal of American Folklore* 43(167): 1–58.

Parsons, E. W. C., 1996. *Pueblo Indian Religion*. (2 vols.) Lincoln: University of Nebraska Press.

Pearsall, D. M., 2000. Paleoethnobotany: A handbook of procedures. Academic Press, San Diego.

Pearsall, D. M., 2019. Case studies in paleoethnobotany: Understanding ancient lifeways through the study of phytoliths, starch, macroremains, and pollen. Routledge, New York.

Powers, S., 1882. North American Indian Legends and Fables. *The Folk-Lore Record*, 5: 93–143.

Powers, S. and Powell, J. W., 1877. *Contributions to North American Ethnology, Volume III: Tribes of California*. Government Printing Office.

Powers, S., and Powell, J. W., 1877. *Tribes of California*. Washington, D.C.: Government Printing Office.

Radin, P., 1926. Wappo Texts First Series. *University of California Publications in American Archaeology and Ethnology*. Berkeley: The University of California Press.

Renaud, E. B., 1925. Archéologie du Sud-Ouest américain. *Journal de la Societé* de américaniste Nouvelle Série, 17: 338–240.

Renaud, E. B., 1930. *Prehistoric Cultures of the Cimarron Valley, Northeastern New Mexico and Western Oklahoma*. Colorado Scientific Society, Denver.

Renaud, E. B., 1942. *Indian Stone Enclosures of Colorado and New Mexico*. University of Denver, Department of Anthropology.

Renaud, E. B., and J. Chatin, 1943. Archaeological Sites of the Cuchara Drainage, Southern Colorado. Archaeological Series Fourth Paper. University of Denver Department of Archaeology, Denver.

Rexroth, A. N., 2010. Prehistoric Sandals of the Southern High Plains: Indicators of Cultural Affinity and Change.

Rhodes, W., 1948. Music of the Sioux and the Navaho. In *Ethnic Folkways Library FE 44001*. Folkways Records and Service Corp, New York.

Rogers, B. T., and Gayton, A. H., 1944. Twenty-seven Chukchansi Yokuts Myths. *The Journal of American Folklore*, 57(225): 190–207.

Russell, Frank. 1898. Myths of the Jicarilla Apaches. *The Journal of American Folklore* 11(43): 253–271.

Sapir, E., and Dixon, R. B., 1910. Yana Texts. *University of California Publications in American Archaeology and Ethnology*, 1–235.

Scheiber, Laura L., and Clark, Bonnie J., 2008. *Archaeological Landscapes on the High Plains*. The University of Colorado Press, Boulder.

Schiavitti, Vincent W., L. L. Loendorf, and Erica Hill, 2001. Archaeological Investigations at Eleven Sites of Welsh Canyon in the Pinon Canyon Maneuver Sites, Las Animas County, Colorado. Midwest Archaeological Center National Parks Service, Lincoln.

Schiffer, M. B., 1983. Toward the Identification of Formation Processes. *American Antiquity* 48(4): 675–706.

Searcy, Michael T. 2011. *The Life-Giving Stone: Ethnoarchaeology of Maya Metates*. University of Arizona Press, Tucson.

Sekaquaptewa, H., and Udall, L., 1969. *Me and Mine: The Life Story of Helen Sekaquaptewa*. Tucson: University of Arizona Press.

Seymour, D. J., 2012. *From the Land of Ever Winter to the American Southwest: Athapaskan Migrations, Mobility, and Ethnogenesis*. University of Utah Press.

Shaw, R. B., S. L. Anderson, K. A. Schulz, and V. E. Diersing. 1989. *Plant Communities, Ecological Checklist, and Species List for the U.S. Army Pinon Canyon Maneuver Site, Colorado*. Colorado State University Department of Range Science, Ft. Collins.

Silverstein, M., 1996. The Secret Life of Texts. In *Natural histories of discourse*, edited by M. Silverstein & G. Urban. Chicago: University of Chicago Press, 81–105.

Simpson, Caryl Wood, 1976. Trinchera Cave: A Rock Shelter in Southeastern Colorado. Unpublished Master's thesis, Department of Anthropology, University of Wyoming, Laramie.

Sparkman, P. S. 1908. Culture of the Luiseño Indians. *University of California Publications in American Archaeology and Ethnology* 8(4): 187–234.

Stansbury, H., 1988 (1852). Exploration of the Valley of the Great Salt Lake. Smithsonian Institution, Washington, D.C. Originally issued as *Exploration and Survey of the Great Salt Lake of Utah, including a Reconnaissance of a New Route through the Rocky Mountains*. Lippincott, Grambo, and Company, Philadelphia.

Stevenson, M. C., 1904. *The Zuni Indians: Their Mythology, Esoteric Fraternities, and Ceremonies*. Washington, D.C.: Government Printing Office.

Stoffle, R. W., 1984. Toyavita Piavuhuru Koroin – Canyon of Mother Earth: Ethnohistory and Native American Religious Concerns in the Fort Carson-Pinon Canyon Maneuver Area: Final Report. University of Wisconsin – Parkside, Kenosha.

Stoler, A. L., 2009. *Along the Archival Grain: Epistemic Anxieties and Colonial Common Sense*. Princeton University Press, Princeton.

Stone, T., 1999. *The Prehistory of Colorado and Adjacent Areas*. Univ of Utah Pr.

Titiev, M., 1948. Two Hopi myths and rites. *The Journal of American Folklore,* 61(239): 31–43.

True, D. L., 1993. Bedrock milling elements as indicators of subsistence and settlement patterns in northern San Diego County, California. *Pacific Coast Archaeological Society Quarterly,* 29: 1–26.

True, D. L., Meighan, C. W., and Crew, H., 1974. *Archaeological Investigations at Molpa, San Diego County, California*. Berkeley: University of California Press.

True, D. L., Pankey, R., and Warren, C. N., 1991. *Tom-Kav: A Late Village Site in Northern San Diego County, California, and its Place in the San Luis Rey Complex.* (Anthropological Records No. 30.) Berkeley: University of California Publications.

True, D. L. and Waugh, G., 1981. Archaeological Investigations in Northern San Diego County, California: Frey Creek. *Journal of California and Great Basin Anthropology,* 3(1), pp. 84–115.

van Gennep, A., 1960. *The Rites of Passage*. Chicago: University of Chicago Press.

van Gennep, A., 2013. *The Rites of passage.* Routledge (originally published in 1909).

Van Royen, W., 1936. The Archaeological Survey of the High Western Plains Seventh Report, Southern Wyoming and SW. South Dakota. E. B. Renaud. (University of Denver, Department of Anthropology, Denver, 1936.) *American Antiquity,* 2(1): 61–63.

Villalobos-Boehm, S., 1997. An Analysis of 201 Bedrock Grinding Features in Utah's Grand Gulch Primitive Area. M.A. Thesis, Department of Anthropology, University of Colorado, Boulder.

von Guerard, Paul, P. O. Abbott, and R. C. Nickless, 1987. *Hydrology of the U.S. Army Pinon Canyon Maneuver Site, Las Animas, Colorado.* Water-Resources Investigation Report 87–4227, prepared by the U.S. Geological Survey in cooperation with the U.S. Department of the Army, Fort Carson Military Reservation, Denver.

Wall, D., and Masayesva, V., 2004. People of the Corn: Teachings in Hopi Traditional Agriculture, Spirituality, and Sustainability. *American Indian Quarterly,* 28(3/4): 435–453.

Wedel, W. R., 1961. *Prehistoric Man on the Great Plains.* University of Oklahoma Presse, Norman.

Wedel, W. R., 1963. The High Plains and their Utilization by the Indian. *American Antiquity* 26(1): 1–16.

White, R. C., 1963. Luiseño Social Organization. *The University of California Publication in American Archaeology and Ethnology* 48(2): 91–194.

Whitehouse, H., 1996. Rites of Terror: Emotion, Metaphor, and Memory in Melanesian Initiation Cults. *Journal of the Royal Anthropological Institute (N.S.),* 2(4): 703–715.

Willey, G. R., and J. A. Sabloff., 1993. *A History of American Archaeology.* W. H. Freeman and Company, New York.

Willis, M. D., Koenig, C. W., Castañeda, A. M., and Black, S. L., 2016. Archaeological 3D mapping: The structure from motion revolution. J. Tex. Arch. and Hist. 3, 1–36.

Wood, Caryl, 1974. Excavations at Trinchera Cave, 1974. Southwestern Lore 40(3–4): 53–56.

Woodbury, R. B., 1954. Prehistoric Stone Implements of Northeastern Arizona. Reports of the Awatovi Expedition. Report No. 6, Cambridge.

Wormington, H. M., 1957. Ancient Man in North America. Denver Museum of Natural History.

Wright, M. K. 1993. Simulated Use of Experimental Maize Grinding Tools from Southwestern Colorado. *Kiva* 58(3): 345–355.

Wyman, L. C., and F. L. Bailey, 1943. Navaho Girl's Puberty Rite. *New Mexico Anthropologist* 6/7(1): 3–12.

Zier, C. J., 2017. The Archaeology of Archaeology: Reconstructing the Excavation History of Trinchera Cave. *Southwestern Lore 83*(2): 21–47.

Zier, C. J., 2019. Comments on the Stratigraphy and Chronology of Trinchera Cave, Southeastern Colorado. *Southwestern Lore, 85*(1).

Zier, C. J., S. M. Kalasz, and M. W. Painter, 1999. *Colorado Prehistory: A Context for the Arkansas River Basin.* Council of Professional Archaeologists, Denver, Colorado.

Index

About the Author

Elizabeth Lynch is currently a post-doctoral researcher with the Hell Gap National Historic Landmark Digital Archives Project at the University of Wyoming in Laramie. She earned her Ph.D. from the University of Wyoming, her M.A. in Biological Anthropology from New Mexico State University, and her Bachelor's in French Language from the University of Texas at Austin. Dr. Lynch has been working with the bedrock ground stone features on the Chaquaqua Plateau since 2007 and has participated in two interdisciplinary symposia on bedrock spaces and landscape in language. Her work focuses on exploring how 3D models and virtual reality are used to visualize scientific problems that evade current methodologies. She is currently developing the use of virtual reality landscape tours for consulting with descendant communities. She teaches classes in anthropology at the University of Wyoming and the University of Northern Colorado.

www.ingramcontent.com/pod-product-compliance
Ingram Content Group UK Ltd.
Pitfield, Milton Keynes, MK11 3LW, UK
UKHW020841290926
15470UKWH00004B/1474

9 781793 618924